PROCEEDINGS OF
THE INSTITUTION OF
CIVIL ENGINEERS

The Channel Tunnel

SUPPLEMENT   TO   CIVIL   ENGINEERING
VOL. 102 · SPECIAL ISSUE 1 · 1994

C O N T E N T S

Part 3: French Section

Channel Tunnel French Section Editorial Panel
C.O. Frederick, Summerfields, Dalbury Lees, Derby, UK

T. W. Mellors, Brylls, West End, Waltham St Lawrence, Reading, UK

A. Thompson, Thomas Telford Publications, Heron Quay, London, UK

Secretary   Lesley Wilson

Editor   Simon Fullalove

Production Editors   Becky Norris, Andrea Platts and Bill Pope

Corrigendum: Channel Tunnel Part 2, Fig. 4, page 25 – the caption should read Holywell Combe, November 1992 and not November 1993.

Written discussion closes 15 September 1994

Channel Tunnel project: Beussingue Portal

*Proc. Instn
Civ. Engrs
Civ. Engng,
Channel Tunnel,
Part 3: French Section
1994, 2*

The Authors

The papers in this Special Issue (originally published in French in *Travaux*, the official journal of the Fédération Nationale des Travaux Publics) were jointly written by the following members of the GIE Trans-anche Construction team.

Henri Barthes

André Bordas

D. Bouillot

Michel Buzon

Phillippe Dumont

Jacques Fermin

Jean-Claude Landry

Jean-Pierre Larive

Laurent Leblond

Jean-Jacques Morlot

Leon Szypura

Phillippe Vandebrouck

Bertrand Vielliard

Note

Throughout this issue chainage is given in km.
The prefix PK = chainage from France.
masl = metres above sea level (taken from the French standard rather than m OD).

TML France: the civil engineering work for the Channel Tunnel in France

Proc. Instn Civ. Engrs Civ. Engng, Channel Tunnel, Part 3: French Section 1994, 3–5

The construction director for the French section of the £10 billion Channel Tunnel project was **Pierre Matheron**, a former project director with SGE. He introduces this Special Issue by describing how the French construction work was organized, how the decentralized management met its objectives and how effective communication with up to 4000 site staff was maintaned throughout the six-year construction programme.

Let us briefly consider the works constructed on the French side of the Channel Tunnel. These consisted of

(a) more than 64 km of tunnels bored using five sophisticated, purpose-made tunnelling machines operating simultaneously

(b) a concrete segment prefabrication factory for the tunnel lining, specially built on site to produce 250 000 tunnel segments at the rate of 9000 per month

(c) a terminal the size of an international airport (700 ha), the function of which is to load into the shuttles road vehicles equivalent to the circulation on a motorway (3450 vehicles an hour).

In addition to the technical challenges each of these elements represented, the most important feature of the project was the very tight programme which led to significant overlapping of the design, procurement, civil engineering and electrical and mechanical installation activities.

Construction organization

The French construction director was responsible to the project executive for all construction activities in France. In particular he was responsible for the budget, complying with the work programmes, and the management of all personnel in France. He supervised the activities of all the sub-projects and the French construction functional department directors over whom he had total authority.

Sub-project operations

During the tunnel-boring phase, operations were organized into four sub-projects

(a) tunnels
(b) prefabrication
(c) terminal
(d) electrical and mechanical installations.

After this first phase, which saw the end of the prefabrication of the tunnel lining segments, and for which the factory only remained open for the manufacture of the tunnel track sleeper blocks and the increase in the installation activities of the electrical and mechanical systems, the project was reorganized into two sub-projects—tunnels and the terminal—which included civil as well as electrical and mechanical works. Each of these sub-projects was placed under the joint responsibility of a sub-project director for the construction works and an engineering director for the design.

The sub-project construction director and the engineering director were jointly responsible for the technical, programme and performance aspects of their sub-project.

Each sub-project construction director

(a) defined the organization of the sub-project, and set up and managed his staff and labour force
(b) managed relations with the client
(c) managed productivity for the sub-project and was responsible for the technical choices for construction as well as programme productivity and establishing and complying with objectives
(d) decided the choice of sub-contractors and suppliers, awarded contracts and managed the various sub-contractors
(e) provided the administration and financial control for the sub-project.

Each engineering director was responsible for the design and optimization of his sub-project.

The sub-projects were subdivided into sections. Normally each work section was headed by a section manager who was responsible for cost control, programme and the choice of technical solutions for construction and, consequently, complying with objectives. It is worth noting that the level of responsibility of a construction section manager varied from Ffr.50 million to Ffr.500 million; these amounts would normally be considered in the public sector as constituting a large project.

The functional departments provided, on behalf of the construction sections and under the responsibility of a department director, common services to the sub-projects, such as administrative, financial, commercial and technical services.

Pierre Matheron, Director, French Construction, GIE Transmanche Construction

Functional departments

There were five functional departments

(a) administration/finance
(b) commercial
(c) project services
(d) human resources and communications
(e) quality assurance.

Generally, the responsibilities of the functional departments were

(a) to define the procedures for French construction, prepared in accordance with TML's general procedures, and to check compliance with the procedures by the sub-projects
(b) to assist, on request, the sub-projects
(c) to centralize management information from the sub-projects for French construction for internal (management) and external use (sent to the client)
(d) to manage non-decentralized operations for the sub-project.

The functional directors had direct links with their corresponding section or service managers from the sub-projects. Instructions were issued in the form of procedures or by the director to the sub-project concerned. The organization of each functional department was modified for the second phase, as the electrical and mechanical works became more important than the civil engineering work. These departments were integrated during this phase, with the tunnel's sub-project and with the corresponding functional sections for the terminal sub-project.

Management system and objectives

In order to meet the complexity and extent of the works to be carried out—and this in a very short period of time—a decentralized management system was created. This system implied the following objectives for all French construction

(a) loyalty of people to the project and development of their capacity to work together
(b) a rigorous safety policy
(c) quality standards
(d) compliance with the programme
(e) control of costs.

In order to achieve these objectives, a management system based on clearly defined rules was set up in the following areas.

Human resources

The fundamental objectives for the personnel were professionalism, safety, employment, quality, information and communication.

In terms of employment, GIE Transmanche was committed to recruiting 75% of its labour locally from the north Pas-de-Calais region. Through a voluntary policy and collaboration with the Government and regional authorities,

this figure has in fact reached 95%. The local recruitment involved setting up special training schemes with all involved parties.

As for the staff, of which approximately half came from the French parent companies making up GIE Transmanche, a management participation structure was set up based on the open flow of information and establishing objectives at section levels.

Safety

A specialized service, in continuous contact with the technical and operational services, prepared the safety procedures. A strict accident prevention policy was created. The policy included an initial training course for all personnel on arrival, with complementary specialized training courses later (first-aid, national certificate of life-saving and resuscitation, electrical awareness, machines, lifting appliances), continuous 24 h presence of a professional fireman and nurse and two full-time doctors employed on the site.

Quality

A management system for quality, placed under the responsibility of specialists who had acquired extensive experience in this field on large construction projects, was set up. This provided a consensus of opinion on the advantages of strict quality management, not only for programming and costs, but also for good client relations. The quality system also trained and raised the awareness of all personnel. In the final event each person was responsible for the quality of production.

Programme

The project programme was divided in a similar manner to that of the organization and included the following levels

(a) Level 1: general programme, managed by the coordination structure
(b) Level 2: sub-project programme
(c) Level 3: section programme.

The periods allowed for each programme were managed by setting objectives at section levels which were consolidated at sub-project level. Level 3 programmes were prepared with the aid of computers and identified activities, interface and required resources. This level provided charts for progress, personnel needs and accounts.

Level 2 programmes included both the design and installation programmes for each sub-project. Lastly, the level 1 general programme was made up of the activities from the two sub-projects.

Costs

Each sub-project was an independent profit centre. It controlled and managed its own costs and receipts

(*a*) by establishing and monitoring financial objectives

(*b*) by preparing a cost end-forecast twice a year

(*c*) by reviewing the cost end-forecast every three months.

The consolidation of the cost control activities and the preparation of reports were carried out by the project services department with assistance from the administrative and financial departments.

Communication within the French construction operations

The communication policy set-up had to meet the peculiar character of this large project, including the

(*a*) varied sources of personnel

(*b*) work regime (three shifts, seven days a week)

(*c*) size of the site

(*d*) location of the personnel (Paris, Calais, Sangatte, Coquelles and Great Britain)

(*e*) necessity of agreeing common policies, and a good knowledge of the project and its progress.

This policy centred along two main lines.

Meetings

The group directorate committee and the sub-projects committees met once a week. This provided better information and helped decision making. Periodic information meetings were held for the whole work force and staff.

Publications and audiovisual media

Two publications, one fortnightly (*L'ouvrage*) and the other monthly, and the Franco-British *The Link* provided news on the progress of the project to all personnel. Every morning, messages on television screens gave details of daily activities on site to all offices.

An internal video network throughout the site provided an instant means of displaying messages, films or clips on work progress, safety, training, and hygiene. Moreover, a news video, transmitted monthly at first but then weekly, provided pictures from the daily life of the site to everyone.

*Proc. Instn
Civ. Engrs
Civ. Engng,
Channel Tunnel,
Part 3:
French Section,
1994, 6–10*

Paper 10486

Tunnels—geology

*H. Barthes, A. Bordas, D. Bouillot, M. Buzon, Ph. Dumont, J. Fermin,
J.-C. Landry, J.-P. Larive, L. Leblond, J.-J. Morlot, L. Szypura,
Ph. Vandebrouck and B. Vielliard*

■ **In 1882, the geology of the Pas-de-Calais region was investigated in connection with a Channel crossing project. After an historical review of these surveys, this Paper describes subsequent exploratory work carried out after the submission and study by means of the geostatic method. It then looks into the geological data, stressing the presence of Chalk Marl which was the subject of additional studies. A description of the geotechnical characteristics and the tectonics of the site is followed by a section on the choice of the alignment based upon these data. The Paper concludes with a detailed description of the geological follow-up studies carried out during the works.**

History

Surveys before 1985
The geology of the Pas-de-Calais has been studied ever since a tunnel was first considered, when the sea-bed was mapped from samples taken during dives (by Thomé de Gamond) or by soundings.

2. Between 1882 and 1883 the actual tunnelling conditions were first investigated when two shafts and an adit 2·10 m in diameter and 1669 m long were excavated at Sangatte under the supervision of the engineer, C. Breton.

3. In 1958–59, in a renewed effort, the 1883 adit was pumped out and the accessible section examined (the furthermost section had been sealed off when the project was abandoned), eight marine boreholes were drilled and a seismic reflection survey was carried out.

4. In 1964–65, an initiative by the Channel Tunnel Study Group gave rise to the most important investigation so far with a seismic reflection survey and 62 marine boreholes in the French section of the Channel and nine land boreholes between Sangatte and Frethun. There followed a programme of laboratory tests on the samples from these boreholes.

5. Between 1972 and 1975 further studies were undertaken by the Situmer group, in the form of three new geophysical surveys, eight marine and seven land boreholes, in order to confirm the results of the 1964–65 investigations. The original shaft at Sangatte was pumped out, two experimental adits driven, and several tests made to determine the ground characteristics.

6. In 1985 the submission to the two Governments was based on the information obtained from surveys to that date. For this submission, three reports were prepared by BRGM, Coyne & Bellier and Simecsol, which abstracted the available data and assessed the anticipated geological and geotechnical problems.

Post-submission investigations
7. Following the submission, complementary surveys were carried out by Simecsol, Wimpol, Geocean, Mecasol and Cotesol. These involved ten cored holes and two pressuremeter tests to investigate the region of the Sangatte shafts, a full seismic reflection survey in the Straits of Dover, and eleven marine and five land boreholes, dilatometer tests, and laboratory tests on samples which sought, in particular, the deformation characteristics of the sub-soils. Two boreholes were sunk close to the site intended for the French crossover (Fig. 1).

8. The geotechnical assessment of all these surveys was undertaken by Simecsol and Mecasol and the geological assessment by BRGM. Results of the latter were analysed by the geostatic method. This allowed spot levels to be predicted from which the base of the Glauconitic Marl could be estimated with reasonable accuracy. The method was also used to study permeability.

Fig. 1. Barge for marine boreholes

Geological and geotechnical data

Stratigraphy and lithology

9. The project encountered strata which comprise, in ascending order:

(*a*) Aptian Greensand

(*b*) Albian, represented by the Gault Clay — dark grey, with phosphate nodules, more calcareous towards the top, 10–13 m thick at the French coast, increasing to over 40 m thick on the UK side

(*c*) Cenomanian (Lower Chalk) often subdivided into the Glauconitic Marl at the base, the Chalk Marl, the Grey Chalk and the White Chalk at the top

  (i) the Glauconitic Marl—greenish and of variable facies, partly clayey, partly sandy, even gritty

  (ii) the Chalk Marl comprises vertical cycles, each of 30–50 cm thick—it is fairly clayey (about 60–80% $CaCO_3$), coloured blue-grey by fine grains of iron sulphide (the upper boundary of this horizon is described in detail below)

  (iii) the Grey Chalk also comprises cycles, but these are 1–2 m thick—overall it has a higher chalk content (generally more than 80% $CaCO_3$)

  (iv) the White Chalk, with no cyclical sediments, is pure white with a $CaCO_3$ content above 85%

(*d*) Turonian, subdivided in the following ascending order:

  (i) a set of chalks very similar to the White Chalk at the top of the Cenomanian, and including a nodular chalk (consisting of chalk pebbles cemented by a greenish type of marl) below marly chalk and nodular beds

  (ii) a set of chalks virtually identical to the Coniacian chalk it underlies, differing from it principally in the smaller number of flints found there—this chalk is distinguished by its high porosity (its void space represents 30–40% of its volume) and its high $CaCO_3$ content (over 90%)

(*e*) Coniacian, known locally as 'Champagne chalk'—white, porous and which can be used for drawing, this chalk forms the hills which lie to the west of Sangatte and can be seen in the cutting at the tunnel portal at Beussingue. It contains flint beds at 0·5–1 m intervals.

Geotechnical characteristics

10. These influenced the design of all the tunnel works, i.e. the tunnel proper, in particular the precast reinforced concrete lining segments, the piston relief ducts and communication passages, electrical rooms, the pumping station and the crossover.

11. Table 1 shows that the coefficient of permeability of the Grey Chalk is between two and ten times higher than that of the Chalk Marl. As seepage in the tunnel is directly proportional to this coefficient, it was important to be able to predict in which formation the tunnel would lie.

12. The upper boundary of the Chalk Marl was difficult to define, and it was therefore essential to find one or more objective criteria by which to identify it. In 1987, based on the work of Amedro and Robaszinsky on the cliff at Cap Blanc-Nez,[1–3] and the work of Carter in 1965, BRGM proposed that a microfossil *Rotalipora reicheli*, found at Cap Blanc-Nez very close to the top of the Chalk Marl, should be the index.

13. The structure of the zone (dip, folds, faults) at the level of the strata through which the tunnels pass is the result of movements which took place in the Tertiary period when the Alps were being formed. The zone lies, however, at the extreme edge of these influences, demonstrated by a northerly dip of 15–20° on the French side, which decreases to less than 5° towards the UK. Numerous faults have been found by geophysical investigation or boreholes, especially in the first 3 km on the French side (six faults with a throw of 2–20 m intersect the alignment and another two faults close by), whereas they are rare on the UK side.

Choice of alignment

14. The alignment is the result of a compromise (Fig. 2). It was desirable to contain the works as far as possible in the stratum where excavation would present least difficulty. The objective was to work in the Chalk Marl, maintaining as far as possible a vertical clearance of 5 m above the top of the Gault Clay and 5 m below the base of the Grey Chalk.

15. The tunnel geometry was also dictated by a maximum gradient determined by the engine power of the trains, a minimum gradient to meet drainage requirements, and minimum radii of the horizontal and vertical curves in relation to train speeds.

16. It was necessary to avoid existing boreholes; in the marine section records showed that some boreholes sunk in 1958 had not been charted and there was also no guarantee of the quality of the backfilling of most of the earlier boreholes. It was therefore imperative to keep clear of them to avoid major inundation.

17. The result was an alignment never less than 35 m from the theoretical position of old uncharted boreholes, which does not invade the Gault Clay. The protection afforded by a desirable 5 m clearance could not be maintained in five sections. The alignment runs well within the recognized stratum and intersects the faults detected in the geophysical survey more or less at right angles, with the exception of the fault

Table 1. Characteristics of soils adopted for the calculation of structures after surveys

Formation	Weight per saturated volume γh: kN/m³	Resistance to simple compression R_c: MPa	Modulus of instantaneous deformation E_i:* MPa	Coefficient of creep ϕ	Coefficient of permeability† K: m/s
Coniacian	19·5	3	2000	1	10^{-6}
Turonian O and N'	19·5	3	2000	1	10^{-6}
Turonian L and M'	23	6	1500	1	3×10^{-6}
White Chalk	21·5				10^{-6}
Grey Chalk PK 0·2–1	22·5	9	2000	1·5	10^{-6}
PK 1–8·5	22·5	9	1500	1·5	5×10^{-6}
PK 8·5–16·3	22·5	9	1300	1·5	$1–5 \times 10^{-6}$
Chalk Marl‡ PK 0·2–1	23·5	9	1500	1·5	5×10^{-7}
PK 1–8·5	23·5	9	1500	1·5	5×10^{-7}
PK 8·5–16·3	22·5–23	6–9	900–1300	1·5	$1–2 \times 10^{-7}$
Crossover§	23	9–12	2300–4000	17	3×10^{-6} – 3×10^{-7}
Glauconitic Marl	23·5	5	900	1·5	5×10^{-7}
Gault Clay	21·5–22	1·8	200	2·1	10^{-9}

* E_i is corrected to the very long term modulus by $E_i = E(1 + \phi)$.

† The chalks are practically impermeable in bulk form. Water can only circulate through fissures and fractures, hence the importance of this variable.

‡ In the Chalk Marl, variations due to the greater presence of marl from top to bottom and from France towards England make it possible to ascertain the properties of the layer depending on the distance from the French coast.

§ A special survey was carried out for the crossover—180 m long and 26 m high (the size of an underground station)—which had to be sited in a horizon 35 m thick and on an incline.

at a point 7·2 km seaward from the Sangatte shaft (PK 7.2), which was met at an oblique angle. Starting from the Sangatte shaft, driving in the Grey Chalk could only be avoided in four of the sections.

Geological procedures during operations

Geological reporting

18. A team of up to five geologists was attached to Transmanche Link (TML); numbers varied according to the amount of work in hand. One was allocated specifically to analysis of the micropalaeontology with the assistance of Monsieur F. Amedro, a consultant whose work on the cliff of Cap Blanc-Nez formed the basic stratigraphic scale used.[1-3]

19. This team prepared geological reports at the face in each tunnel during maintenance breaks, totalling 121 reports from the service tunnel, 83 from the north running tunnel, 70 from the south running tunnel—the equivalent of at least one report per 190 m of driven tunnel.

20. The cross-passages and piston ducts were surveyed, with at least one report from every half-section. Reports were made daily or weekly in the major underground structures— the pumping stations and crossover— depending on the rate of progress.

21. Investigative probe holes were driven in the service tunnels, especially to assess the position of the tunnels in relation to two important boundaries: the top of the Chalk Marl and the base of the Glauconitic Marl (Fig. 3). Rock and water samples were taken for laboratory tests and spoil samples were examined. Rates of water inflow were measured during the drives.

Site investigations

22. *Probe holes at the face.* These were drilled in the first 500 m of the drive and then abandoned because the tunnel-boring machines (TBMs) might have had to operate in closed

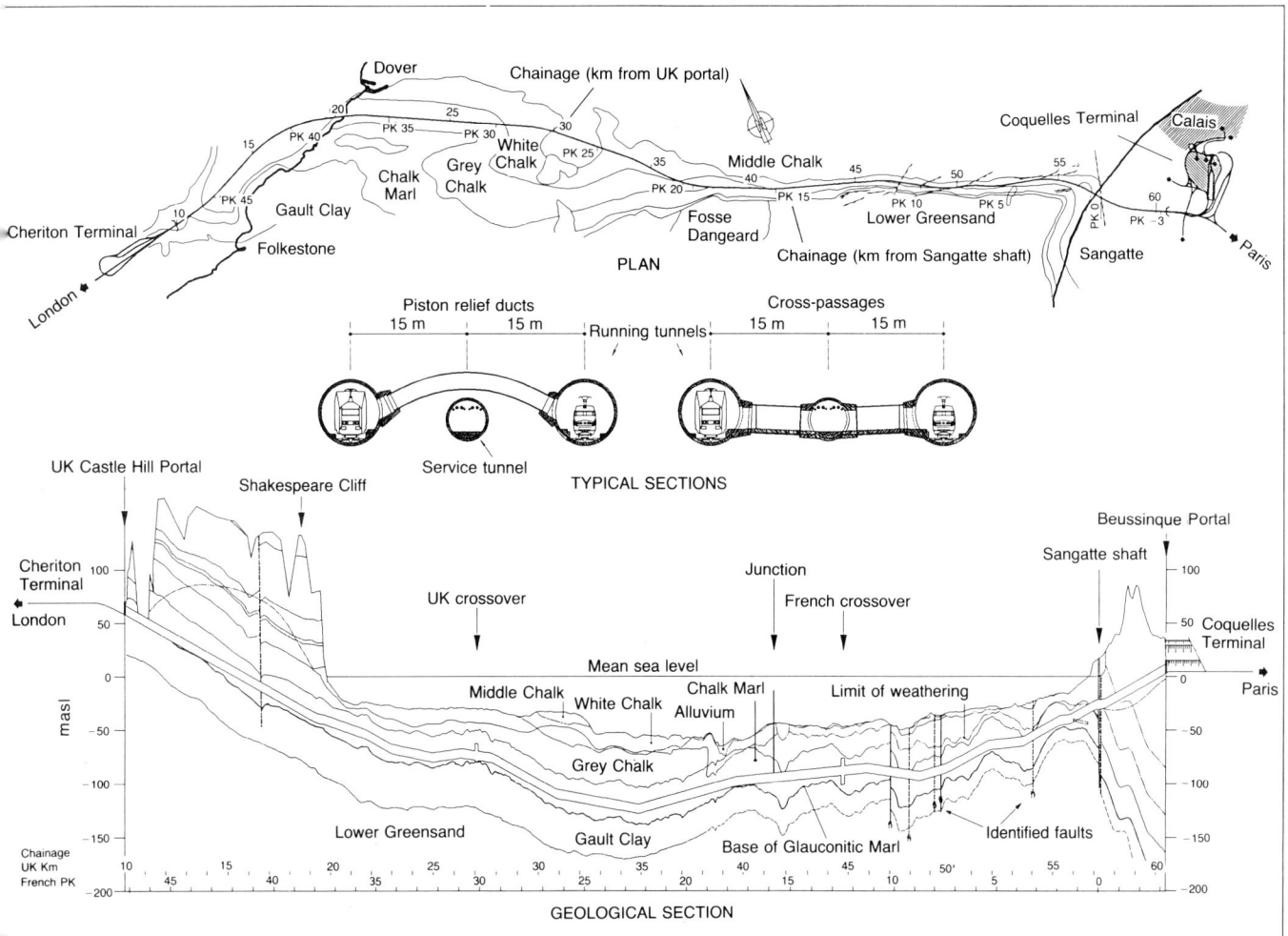

Fig. 2. Horizontal and vertical alignment

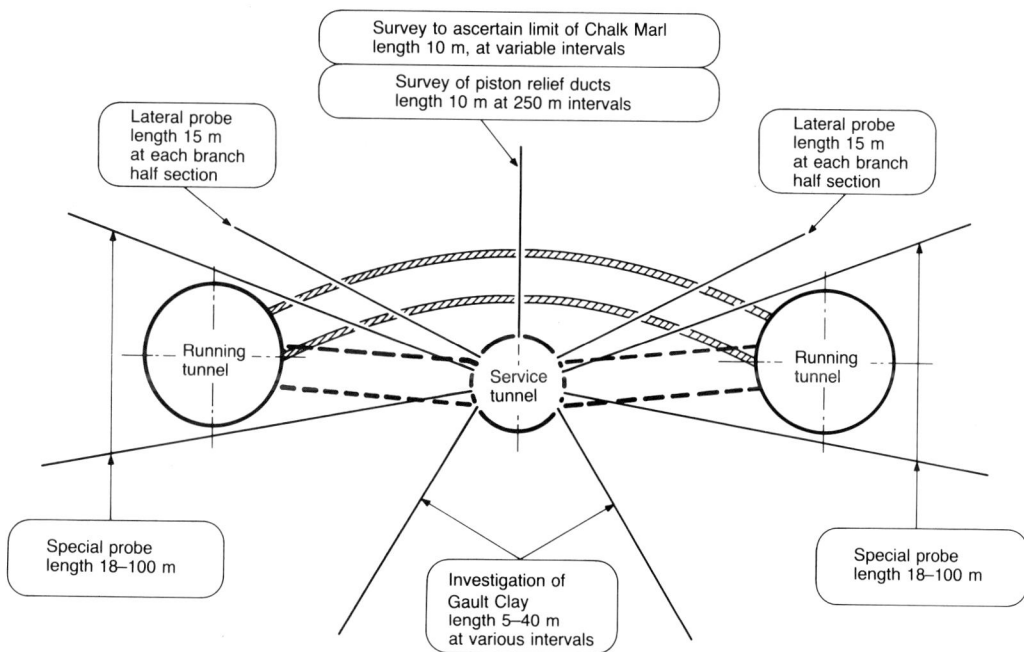

Fig. 3. Reconnaissance works

mode in areas of high water pressure and had to be capable of changing mode quickly.

23. *Boreholes towards the base of the Glauconitic Marl.* Destructive probe holes from drills fixed to a back-up wagon 200 m behind the face were drilled in pairs, one towards the north at about 30° to the vertical, one towards the south at about 40° to the vertical, to verify the angle of dip.

24. *Probes towards the Chalk Marl/Grey Chalk interface.* These probes were executed from a gantry independent of the TBM at each piston relief duct, directed into the crown and often taking core samples. The object was to predict the amount of water likely to be encountered when excavating the piston relief ducts and to check the actual position of the drive in the Chalk Marl.

25. *Special surveys.* Additional boreholes were drilled when local conditions demanded. Close to the fault detected by the geophysical survey at PK 7.6 at a very shallow angle to the alignment, three probes were drilled at right angles to the tunnels to check for major inflows of water. This proved not to be the case; the fault was found to be filled with clay and calcite crystals.

26. Generally predictions of the strata limits were found to be accurate. With regard to the position of the top of the Chalk Marl, a systematic displacement was found, due to a bed missing from the profile of the cliff of Cap Blanc-Nez published in 1980, which served as a reference in the investigations of 1986–87 and 1988. Significant divergences were found at several points on the alignment up to 12 km out from Sangatte where the top of the Chalk Marl was found to be appreciably higher than expected, and locally at 7·8 km where it was estimated to be about 20 m lower. At these points, the nature of the chalk in which the tunnel lay did not change.

27. The spacing of the observations did not generally confirm the presence or absence of faults indicated by the geophysical survey. Only the fault 7·6 km from the shaft was identified by the probes taken from the tunnel. On the other hand, fracturing was observed, particularly when the cross-passages were being excavated. This was particularly significant near the coast.

28. Overall, the permeability of the ground was found to conform reasonably to the values expected. Locally, however, in particular close to the Sangatte shaft, water ingress was very heavy.

29. Investigations carried out during the works confirmed that, in general, ground conditions were as predicted.

References

1. AMEDRO F., DEJONGH E., et al, Les falaises crayeuses du Boulonnais: lithostratigraphie et repères biostratigraphiques de l'Aptien au Sénonien, *C. R. Somm. Soc. Géol. France*, 1976, **3**, 91–94.
2. AMEDRO F., et al, Echelles biostratigraphiques dans le Cénomanien du Boulonnais (macro-micro-nannofossiles), *Géologie méditerranéenne*, 1978, **5**, 5–18.
3. ROBASZYNSKI F. and AMEDRO F., Synthèse biostratigraphique de l'Aptien au Santonien du Boulonnais à partir de sept groupes paléontologiques: foraminifères nannoplancton, dinoflagelles et macrofaunes, *Revue de Micropal*, 1980, **22**, 195–321.

Tunnels — transport logistics

H. Barthes, A. Bordas, D. Bouillot, M. Buzon, Ph. Dumont, J. Fermin,
J.-C. Landry, J.-P. Larive, L. Leblond, J.-J. Morlot, L. Szypura,
Ph. Vandebrouck and B. Vielliard

*Proc. Instn
Civ. Engrs
Civ. Engng,
Channel Tunnel,
Part 3:
French Section,
1994, 11–14*

Paper 10488

■ **The Paper opens with a description of the constraints imposed by a tunnel project. The organization of train deliveries using rails in the tunnels, for the transport of personnel as well as equipment is detailed, and the techniques employed in the removal of spoil, including pumping, are outlined in the final paragraphs.**

General principles

The tunnel works, apart from their length, were significant in that besides the main tunnels themselves there was a series of ancillary works such as cross-passages, piston relief ducts, electrical service rooms off the running tunnels, ground investigation and grouting from the service tunnel, the crossover cavern, the pumping station, and installation of permanent equipment.

2. Transport to the sites in the tunnels was by rail. In the marine workings, service points were set up at approximately 1 km intervals in the service as well as the running tunnels. These were intended for the relay equipment serving the ventilation, pumping and electricity supplies. These points were used for siting the pairs of crossovers on the double-track 900 mm gauge construction railway in each marine tunnel.

3. Open muck wagons were employed for the removal of spoil.

4. The Sangatte shaft was the nerve centre of the logistics of the system since, apart from spoil removal from all the work faces, it was the delivery point for all materials (Fig. 1) and services (ventilation, drainage, electricity, communication systems, compressed air, potable and non-potable water supplies) as well as the transport terminal for personnel, maintenance equipment and emergency services.

Rolling stock

Principal trains serving the TBMs

5. The wagons carrying supplies to the face and the muck wagons used for spoil removal were assembled as a single train to provide all services needed for the installation of one ring. Each main train consisted of, from rear to front,

(*a*) material cars carrying rails, piping, cables, and various supplies
(*b*) one flatbed pallet carrier for the six segments which make up a ring
(*c*) one car with the annular grout for one ring (4·2 m^3 for the marine service tunnel; 2·2 m^3 for the land service tunnel; 5·3 m^3 for the land running tunnels and 8·5 m^3 for the marine running tunnels)

Fig. 1. Storage of segments on palettes at top of shaft

Fig. 2. Transport of concrete

Fig. 3. Personnel boarding a manrider

(d) six to eight muck wagons for the service tunnel, with a nominal capacity of 10·2 m³, or 9 m³ useful capacity with a potential bulking factor of 2

(e) 12 or 13 muck wagons in the case of the running tunnels, of a nominal and useful capacity of 14·4 m³ and a maximum bulking factor of 1·95. The first three wagons were fitted with brakes.

Trains for Special Works and grouting

6. These trains were adapted to the requirements for execution of the Special Works for spoil, concreting (Fig. 2), grouting and investigations using the same battery locomotives as the main trains. The average frequency of the

service was four trains per shift for all the marine running tunnels.

Manriders

7. For freedom of operation and flexibility these cars (Fig. 3) were fitted with diesel engines and operated singly or in threes or fours. Their purpose was to

(a) convey emergency services to the face — one fire-fighting car, one ambulance car, and one mixed service car

(b) transport personnel (eleven manriders) each carrying 13 to 75 passengers

(c) provide maintenance — two electrical service cars, two mechanical service cars, two rail-mounted cranes and one mobile pumping service.

Traction system

8. The main trains and those servicing special works were hauled by two electric locomotives powered by catenaries and batteries (Fig. 4). Each locomotive had a battery charger on board which allowed it to recharge while running.

9. The locomotives at the head and rear of a train had their controls linked by cable which allowed them to be operated by only one driver.

10. Many sections were not equipped with catenaries. These were shunting areas, points, the shaft and marshalling chambers, fixed equipment zones, isolated work zones, the working face zones of the TBMs, and all the land tunnels. In view of this, a catenary installation was not enough to make up all the loco-

*Fig. 4. Combination
trolley/battery
locomotives of
250 kW*

motive needs and it was necessary to have a
recharging station for batteries and means of
periodically exchanging batteries.

11. The frequency of rotation of batteries
depended on the length of the tunnels, on the
length of time the batteries were in use instead
of the catenary supply, allowing for the failure
of the pantographs, frequency of starting and
stopping, and the areas where catenary
supplies were switched off. By the time the

north marine running tunnel had reached the
end of its drive, for example, the batteries were
being exchanged after every two return trips,
taking on average six or seven hours per trip.

Spoil removal

12. There were two aspects to handling the
spoil: the capacity of the handling plant in the
shaft, and the turn-round time of the muck
trains in the tunnels.

*Fig. 5. Spoil removal
control desk*

Fig. 6. Mud pump of 90 m³/h for spoil removal

Capacity of spoil handling plant at the shaft

13.   The facility for extracting spoil by pumping from the bottom of the shaft to the Fond-Pignon discharge site was designed for an advance rate of 3 m/h from four TBMs operating simultaneously (Fig. 5). This was equivalent to 625 m³/h net of chalk in situ, discharged by eight pumps of 90 m³/h capacity (Fig. 6), and with a bulking factor of 1·92 estimated at the time of the initial study.

14.   The average monthly drive rates assumed in the study, excluding start-up times were as follows

(a)   marine service tunnel: 470 m
(b)   marine running tunnels: 540 m
(c)   land running tunnel south: 300 m.

This represents a monthly volume of 95 000 m³ of chalk in situ from the TBMs and about 105 000 m³ in total, allowing for spoil from the crossover and special works.

Turn-round time of spoil trains in tunnels

15.   The maximum authorized speed of these trains was 25 km/h. However, the locomotives for heavy duty convoys were fitted with restrictors to limit their speed to 20 km/h.

16.   The average speed assumed in designing the trains at the time of the study was

(a)   12 km/h for heavy duty convoys
(b)   18 km/h for self-propelled manriders.

In the event, the average speed of the heavy duty convoys was 8 or 9 km/h.

Tunnels—tunnel boring machines

H. Barthes, A. Bordas, D. Bouillot, M. Buzon, Ph. Dumont, J. Fermin,
J.-C. Landry, J.-P. Larive, L. Leblond, J.-J. Morlot, L. Szypura,
Ph. Vandebrouck and B. Vielliard

Proc. Instn
Civ. Engrs
Civ. Engng,
Channel Tunnel,
Part 3:
French Section,
1994, 15–19

Paper 10489

■ **This Paper begins by describing the requirements for the tunnelling machines. Five machines were used, three on the marine tunnels and two on the land tunnels. The description is then developed, distinguishing between the types of machines working under the sea and those under land. A table summarizes the main characteristics of the tunnelling machines.**

Tunnel boring machine design

Most of the French section of the works, like the UK section, lay in a favourable geological horizon—Chalk Marl of considerable thickness and low permeability.

2. However, in contrast to conditions in the UK section, the chalk on the French side had been subjected to greater tectonic movements causing many faults. This meant sections of the tunnel alignment had to run through the much more permeable Grey Chalk, especially in the land sector and in the first kilometre of the marine drives.

3. In order to ensure the safety of the works and to minimize the risks connected with ground treatment, closed-face earth-balance type tunnel boring machines (TBMs) were used, capable of withstanding maximum hydrostatic pressures of 11 bar. A watertight lining was installed with the segments erected from within the tail skin. As such, the TBMs were able to work in water-bearing strata subject to pressure without systematic ground treatment.

4. In the marine section the alignment is through long stretches of low permeability Chalk Marl. The TBMs were therefore also designed to work in open mode at theoretically higher rates of progress.

5. The spoil extraction system was selected with the aim of avoiding those problems connected with long distance liquid transport. This system also allowed excavation at higher speeds when operating in closed mode.

Description of the TBMs

6. The six tunnelling drives were carried out by five TBMs (Table 1). Three machines working simultaneously drove the three marine tunnels: the service tunnel, the north running tunnel and the south running tunnel. Two TBMs drove the three land tunnels in succession: one driving the service tunnel and the other driving the south running tunnel from the Sangatte shaft to the Beussingue Portal and then the north running tunnel from Beussingue back to the shaft.

7. In the marine service tunnel, the ground reconnaissance and grouting units were positioned at the rear of the back-up to treat the zones of excavation of the cross-passages as well as the faulty terrain on the line of the running tunnels. In the land tunnels, only the fissured zones crossed by the TBM in the north

Table 1. Main characteristics of tunnel boring machines

TBMs	Marine service tunnel— Robbins	Marine running tunnels— Robbins Kawasaki	Land service tunnel— Mitsubishi	Land running tunnels— Mitsubishi
Cut diameter: m	5·77	8·78	5·61	8·64
External diameter of shield: m	5·72	8·72	5·59	8·62
Shield length: m	11	13·75	10·60	12
Shield weight: t	470	1250	350	600
Weight of back-up: t	600	800	400	400
Maximum thrust on cutter head: t	1800	2000	4000	9000
Maximum rear thrust: t	4000	11500	4000	9000
Maximum torque: tm	360	1300	400	1300
Installed power: kVA	2700	4000	2500	3800
Cutter head speed: rev/min	2·5 or 5	1·5 or 3	0·9 or 1·8	1 or 2
Maximum advance rate: cm/min	12	12	8	8
Number of back-up wagons	19	16	13	14
Distance from face to last wagon: m	318	265	204	211
Distance from face to end of Californian crossover points: m	368	305	244	251

Fig. 1. Cutter head of a marine running tunnel TBM being mounted at the bottom of the shaft

Fig. 2. Rear shield of a marine running tunnel TBM being lowered into the shaft

excavation in the event of cutter disc failure.

(b) A steerable front body in which the telescopic cutter head moved. This shield was fitted with front and rear thrust rams.

(c) A rear shield with grippers which pushed against the excavated ground, linked to the forward unit by steering rams (Fig. 2).

(d) A spoil extraction system which, in closed mode, permitted the hydrostatic pressure exerted at the front of the machine to be progressively reduced, the spoil being then discharged to a conveyor. This arrangement was applied to the two types of TBM.

    (i) The marine service tunnel TBM was fitted with a continuous screw which fed two piston discharge pumps, isolating the front from the rear independently of the nature of the ground being excavated and of the hydrostatic pressure (Fig. 3). The system worked well in general, except that it did not help to form a watertight and stable plug confining the materials in the cutter head.

    (ii) The marine running tunnel TBMs had an arrangement of two screws in series separated by a compartment of compressed spoil (Fig. 4). Adjustment of the relative speed of the two screws permitted continuous extraction of the confined spoil. This system also allowed the pressure at the face to be well controlled; on the other hand, confining the spoil in the screws was a delicate operation and the high axial stress could mean rapid wear to the screws.

(e) A single or double erector fitted with suction pads which placed the prefabricated segments of the lining inside the tail skin of the rear section of the shield (Fig. 5).

(f) A watertight seal at the rear between the tail skin and the tunnel lining was provided by four rows of wire brush tail seals between the segments and the tail skin. The annular void between the rows of seals was continuously fed with a special purpose mastic.

The land tunnel TBMs

10. The land tunnel TBMs were of simpler design and only operated in closed mode. Each comprised a one-piece shield which carried the cutter head. They advanced by means of rams thrusting against the tunnel lining. The segments were placed by two erector arms. The watertight seal between the tail skin and the lining was provided by three rows of wire brush tail seals between the segments and the tail skin. The annular void between the rows of

land running tunnel were treated from the land service tunnel.

The marine tunnel TBMs

8. The marine tunnel TBMs were dual purpose, being able to work in open or closed mode. The shield body which supported the tunnel bore consisted of two telescopic articulated sections—a forward telescopic cutter-head shield and a rear segment erection and gripper section with a tail skin ensuring watertight sealing against the tunnel lining.

9. The main parts of the shield consisted of the following.

(a) A telescopic cutter head which excavated by rotation (Fig. 1). It was open and had six radial arms (service tunnel) or eight arms (running tunnels) with cutter discs mounted on the centreline of arms and picks on the edges. The picks were recessed behind the discs to protect them, their purpose being to ensure continuous

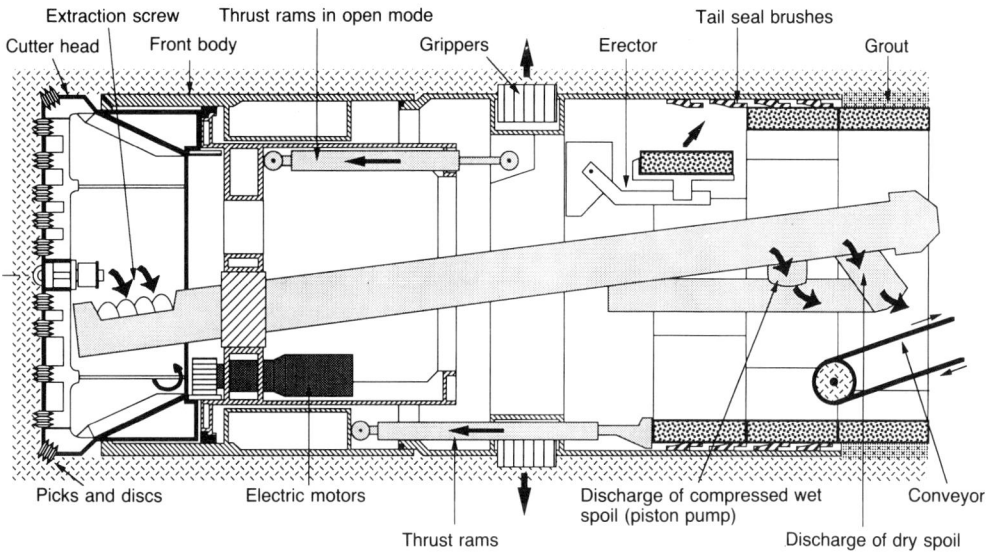

Fig. 3. Marine service tunnel TBM

Labels in Fig. 3: Extraction screw · Thrust rams in open mode · Cutter head · Front body · Grippers · Erector · Tail seal brushes · Grout · Picks and discs · Electric motors · Discharge of compressed wet spoil (piston pump) · Thrust rams · Discharge of dry spoil · Conveyor

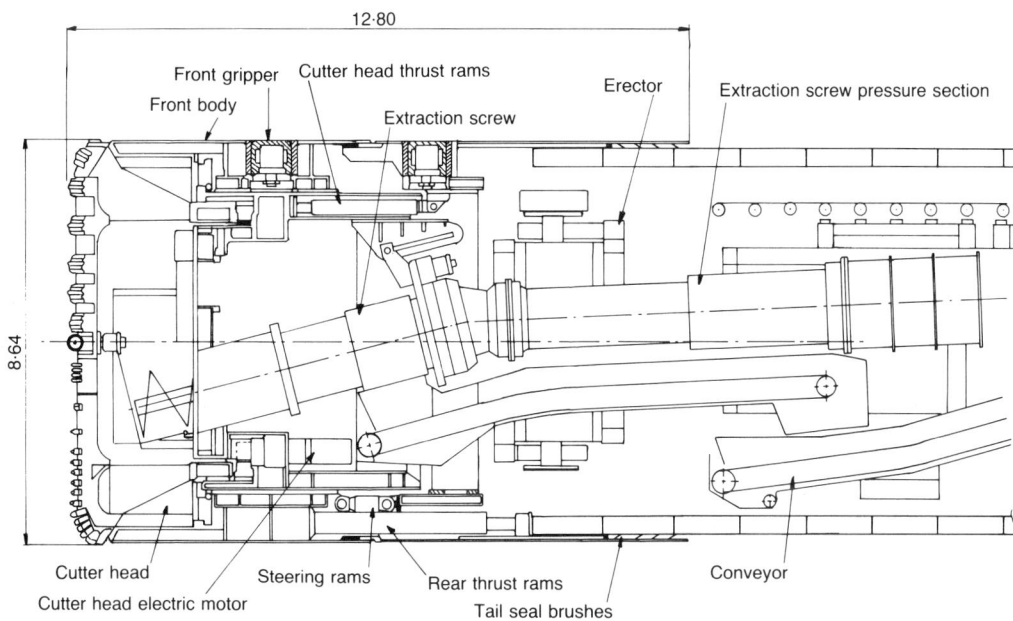

Fig. 4. Marine running tunnel TBM (dimensions in m)

Labels in Fig. 4: 12·80 · Front gripper · Cutter head thrust rams · Erector · Extraction screw pressure section · Front body · Extraction screw · 8·64 · Cutter head · Steering rams · Rear thrust rams · Conveyor · Cutter head electric motor · Tail seal brushes

seals was continuously fed with a special mastic, as for the marine tunnel TBMs.

11. By injecting high density mud in the cutter-head chamber, the land tunnel TBMs could also operate as earth pressure balance (EPB) machines. The spoil extraction system allowed for a maximum hydrostatic pressure of 3 bar.

12. The land service tunnel TBM had a double screw which discharged spoil on to a conveyor belt (Fig. 6). The second screw had an outer casing which ran the spoil in the opposite direction, thus forming a plug of material which maintained a seal.

Fig. 5. Suction device under segment erector of marine service tunnel TBM

10·040

Shield
Cutter head motor
Erector
Tail seal brushes
Extractor screw pressure section

Ø 5·610

Extractor rams

Cutter head
Thrust rams
Roundness gauge
Bearing for first back-up wagon

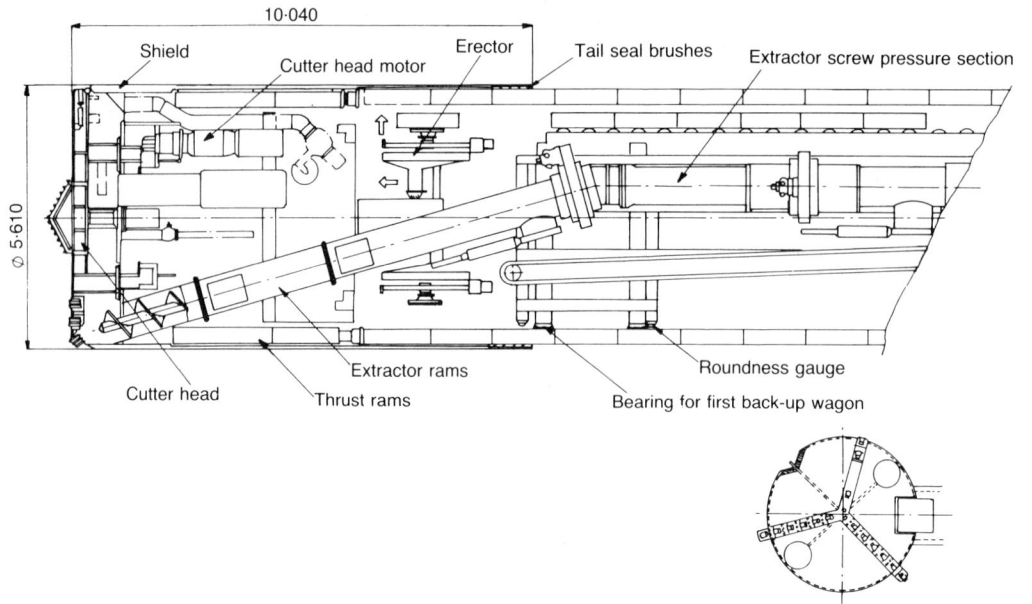

Fig. 6. Land service tunnel TBM (dimensions in m)

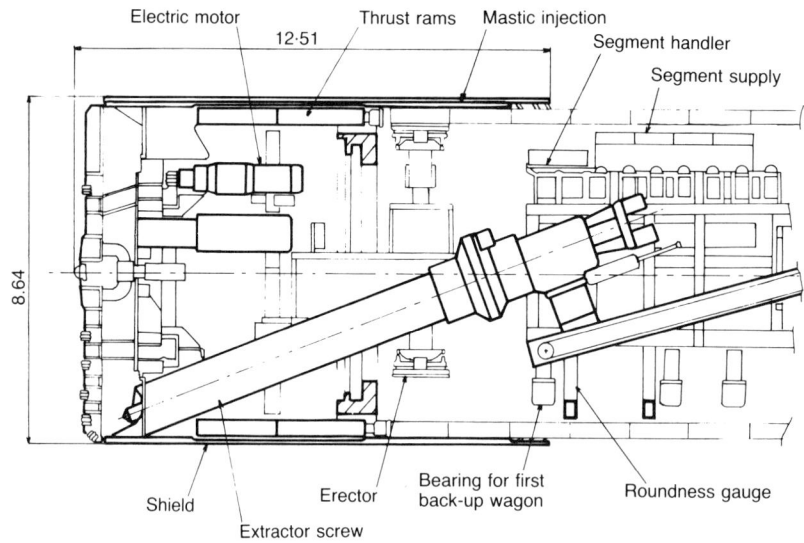

Electric motor
Thrust rams
Mastic injection
Segment handler
Segment supply

12·51

8.64

Shield
Erector
Extractor screw
Bearing for first back-up wagon
Roundness gauge

Fig. 7. Land running tunnel TBM (dimensions in m)

Fig. 8. First back-up wagon for a marine running tunnel TBM

13. The land running tunnel TBM had a single screw (Fig. 7). The spoil was confined solely by regulating the size of the discharge gate feeding the conveyor belt.

The back-up

14. The wagons of the back-up carried the equipment necessary for progress: hydraulic power, electrical supplies, grouting equipment, lubricants, normal and standby drainage, ventilation and dust removal, track laying equipment, pipes, electric cables, compressed air supplies, staff facilities, erection of catenaries in the running tunnels, control cabin, supply and handling of segments, and removal of spoil by conveyor belt (Fig. 8).

15. The number of wagons in each back-up varied between 13 and 19 according to the type of TBM. The length of the back-up from the cutter head to the end of the last wagon varied from 204 m to 318 m.

16. Some 40–50 m further back, beyond the 'Californian' crossover points attached to the rear of the back-up, was the normal double-track arrangement in the tunnel for spoil removal operations and material supply.

17. The originality of the method of moving the first wagon should be noted. The rear of the wagon stood on the track whereas the front was supported by 'shoes' which 'walked' on the newly laid segments about 8 m from the last ring erected.

Proc. Instn
Civ. Engrs
Civ. Engng,
Channel Tunnel,
Part 3:
French Section,
1994, 20–22

Paper 10491

Tunnels—electric power supply

H. Barthes, A. Bordas, D. Bouillot, M. Buzon, Ph. Dumont, J. Fermin,
J.-C. Landry, J.-P. Larive, L. Leblond, J.-J. Morlot, L. Szypura,
Ph. Vandebrouck and B. Vielliard

■ **Almost all the power consumed by the Channel Tunnel project (98%) was electrical. For an installed capacity of 85 000 kW, consumption was that of a town of 25 000 inhabitants. This energy was provided by the Gravelines nuclear power plant, backed where necessary by the Paluel nuclear plant. On the site, electric power was distributed by three ring-main circuits. The main functions on the underground sites were clearly differentiated and the power circuits were doubled to ensure the dependability of the network. Finally, a stand-by power plant using diesel engines was provided to supply 8 500 kW when necessary.**

Electricity accounted for 98% of the power for the Sangatte site—equivalent to the amount of power consumed by a town of 25 000 people. The installed power capacity was 85 000 kW. The nuclear power station at Gravelines, around 40 km to the north of Sangatte, was the source of power which turned the cutter heads of the TBMs and supplied the whole installation.

2. Power for the site was taken from the Electricité de France (EdF) high voltage substation at Les Mandarins, 5 km away. This is one of the most reliable sources of supply in the region, being part of the supply line from France to the UK. Further security was given in the unlikely event of power failure at Gravelines by a back-up supply from Paluel (Seine–Maritime).

3. Two overhead high voltage lines of 90 kV delivered power to the Transmanche Link (TML) main substation at Sangatte, which had two 90/20 kV transformers each of 36 MVA. For normal running, a single high voltage line and one transformer supplied the site; the second line was kept live and the second transformer, although not operating, was in a state of readiness.

4. The installations on site had a total capacity of 90 MVA. If all the equipment were running simultaneously, this would represent a total potential consumption of 60 MVA, of which 6 MVA was channelled to the precasting factory, 4 MVA to the surface installations, 6 MVA to spoil pumping, 8 MVA to the shaft installations, 18 MVA to the marine running tunnels, 8 MVA to the marine service tunnel, 4 MVA to the land tunnels (one TBM at a time) and 4 MVA to emergency pumping equipment.

5. In the main period of tunnel driving (1989 to mid-1991) power consumption exceeded 30 MVA, peaking at 36 MVA in November 1990.

Ring main supply circuits

6. From the 90/20 kV transformer, power was distributed via ring mains to the precasting factory and to substations 1 and 3 (Fig. 1). The remainder of the medium voltage distribution was from substation 2.

7. Substation 1, of 1000 kVA without standby facilities, provided a low voltage supply (380/220 V) to the building attached to the shaft. Similarly substation 3 (1000 kVA) supplied the workshops and the drainage water treatment station. On the same circuit, substation 7 (800 kVA) served the special works workshop and the glue shop, while substation 9 (400 kVA) supplied 750 V direct current to the catenary testing line via a rectifier.

8. Substation 2, installed in the building attached to the shaft, was fed by six lines: two pairs from the 90 kV station and one pair from the emergency supply. The positioning of this substation was important because it provided a medium voltage supply to the tunnels, the five 1600 kVA spoil extraction pumps in the shaft, the two 1600 kVA battery recharging rooms for the locomotives, the floodlights (3·2 kVA) and substations 4, 5, 6 and 8. It also supplied low voltage (three 1600 kVA) to the shaft (ventilation, hoists, lifts, lighting, traverse carriages and tippers, Fig. 2).

9. Substation 4 (1000 kVA) supplied lighting to the various offices on site (except for those attached to the shaft). Substation 5 (1250 kVA) supplied the overhead cranes, substation 6 (1000 kVA) supplied the concrete batching plant and the pumps in the old disused adit, and substation 8 (1000 kVA) supplied the booster pumps for washing the muck wagons.

Underground supplies

10. The underground sites were supplied with medium voltage from substation 2. Each of the main functions had a discrete electricity supply to eliminate practically all risk of general failure other than from a total power cut. In each marine tunnel the following was provided.

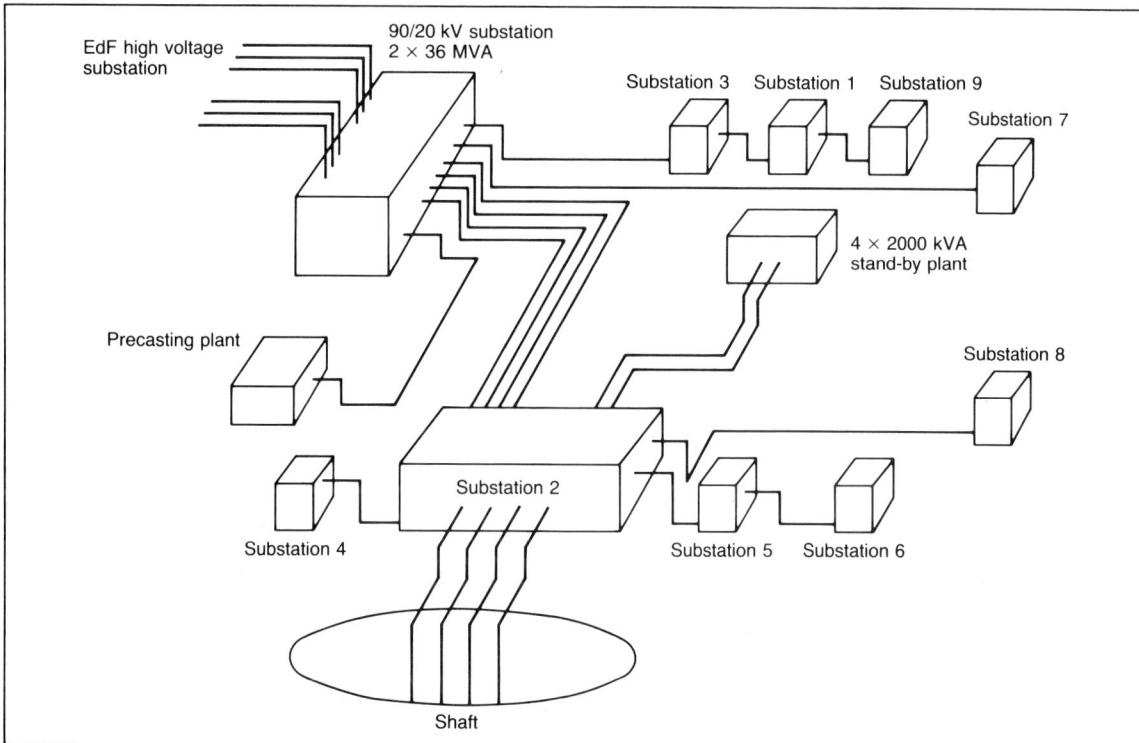

Fig. 1. Surface distribution of high and medium voltage at Sangatte site

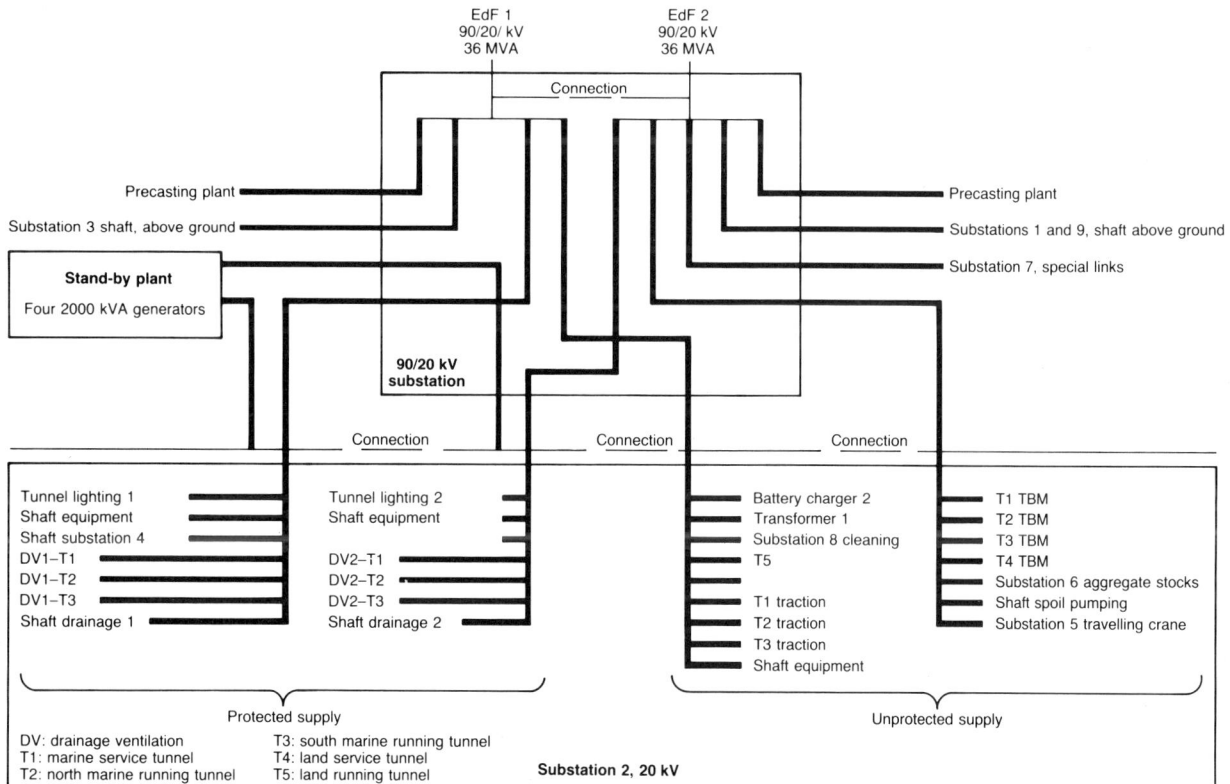

Fig. 2. Substation 2

Fig. 3. Service tunnel showing catenaries for temporary transport system

(a) One 20 kV cable to supply the head of the TBM and the grouting and investigation equipment at the rear of the marine service tunnel machine.

(b) One 20 kV cable to supply the rolling stock (a transformer–rectifier to supply 750 V of direct current to the catenaries installed every 2 km (Fig. 3)). This cable supplied the cross-passage excavation sites.

(c) Two 20 kV cables each kilometre to supply by ring main the substations to boost the ventilation and drainage. They also provided low voltage (380 V three-phase and 220 V single-phase) to the mobile transformer posts at smaller works. This circuit also provided an emergency supply to the TBM via a branch cable at the end of the line.

(d) Two 3·2 kV cables for lighting.

11. In the land tunnels the circuits were simpler because one 20 kV cable for the TBM (also supplying the cross-passages) and two 3·2 kV cables for the lighting were sufficient. There was no need for a cable for the rolling stock, since the locomotives were battery driven, nor for the ventilation which was operated from the shaft. Drainage was by gravity in the service tunnel and the south running tunnel.

12. In the north running tunnel, driven from the terminal, electricity was provided to the TBM by an extension to the south running tunnel supply, and an extension to the land service tunnel cable supplied the installations at the Beussingue Portal for ventilation and spoil discharging. If necessary the second cable could support the first.

Tunnel lighting

13. For the tunnel lighting network, 3200 V/220 V transformers were sited every 330 m. Two independent supply cables each fed every other transformer. For increased safety they were fixed in different parts of the tunnel section.

14. Each transformer supplied 220 V (single-phase) to a circuit of power points. These points had light fittings with two 36 W non-halogen fluorescent tubes. These fittings were spaced to provide the level of lighting required.

Duplicated supply

15. The safety of the electricity network, especially the supply, received careful attention. All the incoming supply installations (transmission to the 90 kV station, 90 kV/20 kV transformers, connection to the shaft, etc.) were duplicated. All automatic control systems were provided with batteries for continuity in the event of power cuts. The drainage, ventilation and lighting systems were all on ring circuits. The marine TBMs benefited from an emergency supply from a branch circuit off the drainage supply line. If the main EdF supply were to fail, the priority networks would be supplied within 30 s from the stand-by generators.

Stand-by generator station

16. Next to substation 2, the stand-by station consisted of four generating sets of 2000 kVA, one of 1400 kVA and one of 315 kVA. The total capacity amounted to 8500 kW, or about 11 500 hp. The sets were tested every week and should have started up automatically 7 s after a power failure. The station fed substation 2 via a duplicated line carrying 20 kV.

17. In normal operation and taking the maximum length of the three tunnels, the stand-by supply was distributed between the marine TBMs (1·4 MVA), ventilation (1·8 MVA), tunnel lighting (0·2 MVA) and permanent drainage (3·1 MVA).

18. In the event of major inundation, work would have stopped and the power would have been diverted to tunnel lighting (0·2 MVA), permanent drainage (2·5 MVA) and, above all, to the emergency drainage equipment (3·1 MVA).

19. Finally, if the stand-by station were to fail to start up, a manually started generator of 315 kVA would supply the priority auxiliaries at the substations and strategic installations like the radio, telephone and central control post. The emergency supply could provide a maximum of 9·7 kVA and, in the event, the most that was required was 6·1 kVA in mid-1991.

Tunnels—the Fond-Pignon discharge site

H. Barthes, A. Bordas, D. Bouillot, M. Buzon, Ph. Dumont, J. Fermin, J.-C. Landry, J.-P. Larive, L. Leblond, J.-J. Morlot, L. Szypura, Ph. Vandebrouck and B. Vielliard

Proc. Instn Civ. Engrs Civ. Engng, Channel Tunnel, Part 3: French Section, 1994, 23–25

Paper 10492

■ **The natural basin of Fond-Pignon, about 2 km from the Sangatte shaft, was chosen as the discharge site for tunnelling spoil. It was therefore necessary to build an embankment across the basin. This Paper describes the embankment and its retaining capacities and discusses the characteristics and consolidation process of the spoil, which was pumped from the shaft as a slurry.**

Site description

The Fond-Pignon basin lies in a depression on Mont-Saint-Hubert next to the site of the Sangatte shaft between 45 and 120 m above sea level (masl) and around 2 km from the shaft access to the tunnels (Fig. 1). The area allocated by the French Government to the fixed link represents about 50 ha and the final area covered with hydraulic fill from the TBMs is 32 ha.

Embankment design

2. The embankment which retains the slurry pumped from the shaft to Fond-Pignon was designed by Coyne & Bellier. It was built of White Chalk fill from five borrow pits in the basin itself (Fig. 2). The three construction stages are detailed in Fig. 3 and Table 1. The total volume of material in the embankment was 1 864 000 m³.

3. The embankment comprises

(a) chalky fill treated with lime and compacted to 100% optimum normal proctor
(b) a main drain, 2·5 m wide, more or less parallel to the inner face, connected to a blanket course and a network of drains— this drainage layout was intended to catch any water seeping though the inner face
(c) three drainage blankets behind the main drain, but not connected to it, acting as drains to consolidate the chalk fill
(d) an overflow system of two groups of spillways along the inner face to evacuate enough water to deal with a 1000-year flood (5 m³/s), decanted water from consolidation of the slurry and rainfall.

Retaining capacity

4. The retaining capacity of the embankment measured at various levels is

(a) $1\cdot4 \times 10^6$ m³ at 66 masl
(b) $3\cdot3 \times 10^6$ m³ at 77 masl
(c) $5\cdot7 \times 10^6$ m³ at 85 masl

When the tunnel drives were completed on 14 July 1991 the spoil was at its maximum height (around 84 masl) for a stored volume of $5\cdot4 \times 10^6$ m³ which represents $3\cdot07 \times 10^6$ m³ of excavated chalk (Fig. 4).

Slurry

5. The grading and water content of the slurry pumped from the shaft changed progressively (Fig. 5). The slurry produced before October 1989 was more granular, 25% passing an 80 μm sieve and with 25% water content. After that date 90% passed the 80 μm sieve and had a water content of 90–100%. It became increasingly clayey after completion of the land

Fig. 1. Location plan of the Sangatte shaft and the Fond-Pignon discharge site

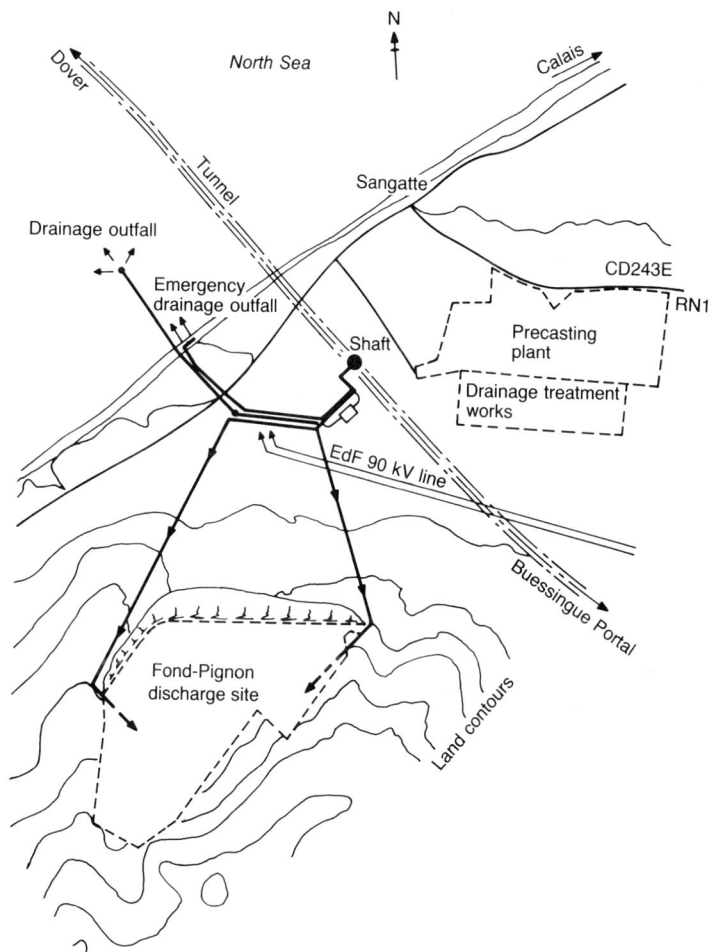

Table 1. *Construction phases of the Fond-Pignon embankment*

Phase	Crest level: masl	Maximum height: m	Gradient of inner face	Gradient of outer face	Crest length: m	Volume of embankment: m³	Dates:
1	67	19	1 in 2·5	1 in 2·5	730	570 000	Aug. 1987– April 1988
2	79	30	1 in 2	1 in 2·1	990	519 000	July 1989– Sept. 1989
3	85·5	37·5	1 in 2	1 in 1·95	1168	775 000	July 1990– Oct. 1990

Fig. 2. *Construction of first phase of Fond-Pignon embankment*

Fig. 3. *Cross-section of the Fond-Pignon embankment showing construction phases*

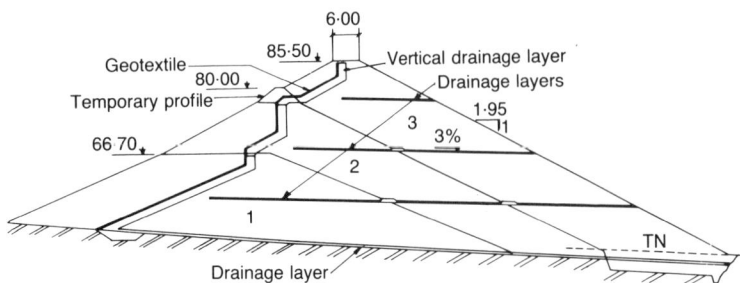

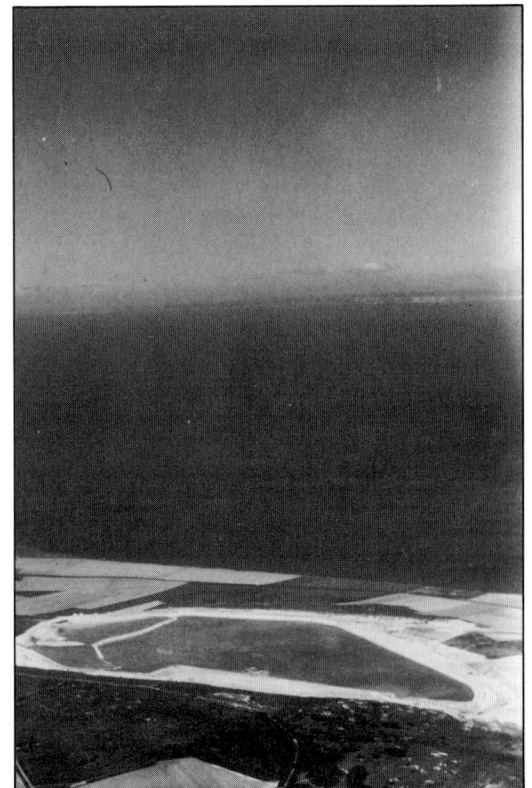

tunnel excavations in the Grey and White Chalk and the increasingly higher percentage of clay in the Chalk Marl in the marine tunnels. The clay content increased from 15% at the bottom of the fill to 25% in the slurry produced by the end of the tunnelling activity.

Bulking of the slurry

6. The term 'bulking factor' applies to the total volume of the slurry produced after handling 1 m³ of excavated chalk.

7. At the time it was dumped the slurry bulked by a factor of about 2. The average bulking factor since slurry dumping started varied from 1·67 before October 1989 to 1·8

after that date. This figure took into account the variations in volume as a result of the reduction in water content when discharging the slurry, or due to consolidation at the bottom of the fill or evaporation.

Consolidation of the slurry

8. Theoretical analysis of the consolidation was carried out by Coyne & Bellier, and also by Mecasol, which gave the degree of consolidation of the slurry both while the discharge site was being filled and on completion.

9. The latter theoretical model was based on undisturbed samples from the centre of the fill and by measuring interstitial pore pressure in the slurry.

Fig. 4. *Fond-Pignon embankment completed*

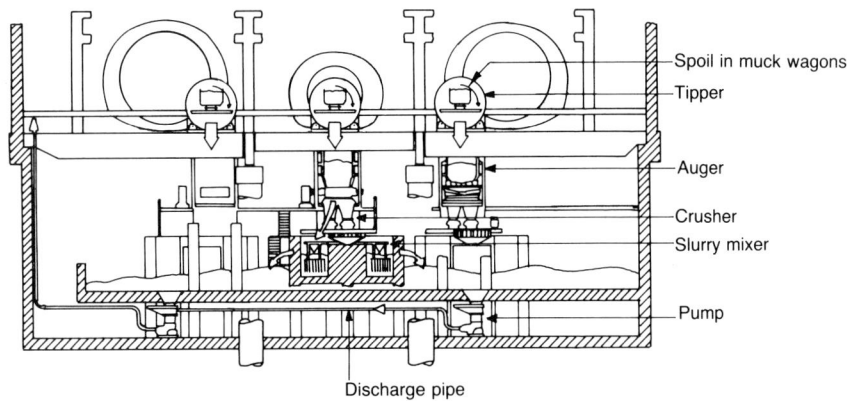

Fig. 5. Spoil handling at the base of the Sangatte shaft

10. The following conclusions were drawn.

(*a*) In a sample of slurry of *H* m in height, the top two-thirds did not consolidate while the discharge site was being filled, whereas the bottom third consolidated well under its own weight.

(*b*) The mean coefficient of consolidation was around 2.5×10^{-7} m²/s, which allowed a more accurate time factor to be included in the consolidation model. The length of time taken by the fill to consolidate under its own weight could also therefore be more accurately obtained. It would take 18 months for the shallow areas and up to six years for the deeper ones to achieve 50% consolidation.

(*c*) The settlement was calculated to be of the order of 4–5 m for shallower areas (15–20 m deep) and 6·5 m for the deeper areas (35–40 m deep).

*Proc. Instn
Civ. Engrs
Civ. Engng,
Channel Tunnel,
Part 3:
French Section,
1994, 26–29*

Paper 10493

Tunnels—dewatering

*H. Barthes, A. Bordas, D. Bouillot, M. Buzon, Ph. Dumont, J. Fermin,
J.-C. Landry, J.-P. Larive, L. Leblond, J.-J. Morlot, L. Szypura,
Ph. Vandebrouck and B. Vielliard*

■ **Three causes of water ingress were taken
into account during the driving phases:
lining leakage rates, normal and accidental water inflows related to the excavation
sites, and the residue of service water used
in tunnels. This Paper reviews these
assumptions and then describes the
dewatering installations at the Sangatte
shaft, in both the land and marine
tunnels.**

Assumptions of discharges

During tunnel driving, water could have
entered the tunnels in three different ways:
seepage from the lining, normal or accidental
inflows connected with excavations (TBMs and
special works) and wastewater from machinery
used in the tunnels.

Seepage from the tunnel lining

2. The acceptable final seepage from the
lining specified in the contract was 1600 m^3/h
in all the marine tunnels. The standard
achieved was considerably higher than the spe-
cification and a rate of 5 l/s per km of tunnel
was maintained during the working period
before breakthrough of the service tunnel. This
accounted for all the seepage, not only from the
tunnels but also from the special works excavated in the bare rock in grouted—and therefore relatively dry—ground. By the time of the
breakthrough of the service tunnel the rate had
been reduced to 2 l/s.

3. Compared with the recorded seepage on
completion of tunnelling, it was possible to
reduce this figure further to 1 l/s per tunnel km,
both at the completion of all the civil engineering work and the installation of mechanical and
electrical services.

Normal water inflows at the face

4. Normal water inflows at the face came
from escapes at the joints in the tail of the
TBMs and were of the following order at
maximum operating water pressure

(*a*) TBM in marine service tunnel: 10 l/s
(*b*) TBMs in both marine running tunnels:
15 l/s each
(*c*) land TBMs: 1 l/s each.

Accidental water inflows via the TBM

5. Accidental water inflows via the TBM
were caused by deteriorating joints in the

TBMs and were of the following order

(*a*) marine: 150 l/s (maximum pressure 11 bar)
(*b*) land: 50 l/s (maximum pressure 3 bar).

Accidental inflows of water in tunnels

6. In the special works, the crossover and
pumping station W3 (excavated by traditional
means), allowance was made for the possibility
of a sudden inundation of 500 l/s. This was the
equivalent of meeting, during excavation, an
old marine borehole of 200 mm diameter which
had been neither charted nor backfilled.

7. If a major inundation occurred in a
tunnel, all excavation would have stopped;
there was thus no need to allow for the possibility of a 150 l/s and 500 l/s inflow accumul-
ating in the same tunnel. If major inundations
occurred in two tunnels simultaneously, all
works would have been stopped.

Discharges

8. In the event, the discharge actually
involved was from the service water supply,
amounting to around 10 l/s per tunnel. The
drainage requirement in the marine tunnels can
be summarized as follows.

(*a*) Permanent drainage
   (i) Seepage in tunnels, including special
     works: 78 l/s (5 l/s per tunnel km over
     15·8 km)
   (ii) Inflows at TBM joints: 15 l/s (10 l/s in
     the service tunnel)
   (iii) Leaks from service water pipes: 10 l/s
     (maximum discharge at any one time)
   (iv) Total maximum per tunnel: 103 l/s
     (98 l/s in the service tunnel)
(*b*) Emergency drainage
   (i) Running tunnels—the design dis-
     charge rate of emergency water was
     150 l/s between the face and the special
     works site nearest to the face; there-
     after 500 l/s to the shaft.
   (ii) Service tunnel—the discharge of emer-
     gency water was 150 l/s between the
     face and the crossover, thereafter
     500 l/s to the shaft.

In the land tunnels the summary is as follows.

(*a*) Permanent drainage
   (i) Seepage water in tunnels: 16 l/s (5 l/s
     per tunnel km over 3.2 km)
   (ii) Inflows at TBM joints: 1 l/s
   (iii) Wastewater: 17 l/s
   Total maximum per tunnel: 34 l/s.

(b) Emergency drainage
 (i) Inflows of water due to deterioration of TBM joint: 50 l/s
 (ii) Accidental inflows in excavating special works: 70 l/s
 (iii) Maximum discharge accounted for: 120 l/s.

Drainage of Sangatte shaft

9. At the bottom of the shaft, two sumps had the capacity to receive a total of 3500 m³ of water from the drainage system of the land and marine tunnels. They were connected by a pipe feeding the four pumps which raised the water to the surface to a height of about 70 m via three 400 mm diameter pipes.

10. A pump with a capacity of 325 l/s ensured permanent drainage in line with the maximum total discharges. When breakthrough of the service tunnel was completed, the two running tunnels had reached a point 15·8 km from the shaft and the three land tunnels were completed. Two similar pumps of 325 l/s

ensured maximum emergency drainage and a fourth, of the same capacity, was kept as a back-up.

11. The water from the permanent drainage systems was discharged into the sea through a diffuser buried near the low tide mark after passing through the 320 l/s treatment plant. This water was also used to add to the muck during the pumping process. The emergency flow was discharged directly into the sea through a diffuser at the foot of the cliff.

Drainage of the land tunnels

12. The service tunnel and the first running tunnel were driven uphill, and water extraction was by gravity flow as far as the sump in the shaft. The second running tunnel was driven after the TBM had been turned around at the Beussingue Portal. Being a downhill drive, it was fitted with two drainage installations— permanent drainage, capable of 40 l/s, and stand-by drainage—giving a total extraction rate of 150 l/s.

Fig. 1. Tunnel drainage Phase 1: (a) from low point 8·7 km from shaft; (b) from face to high point 11·3 km from shaft

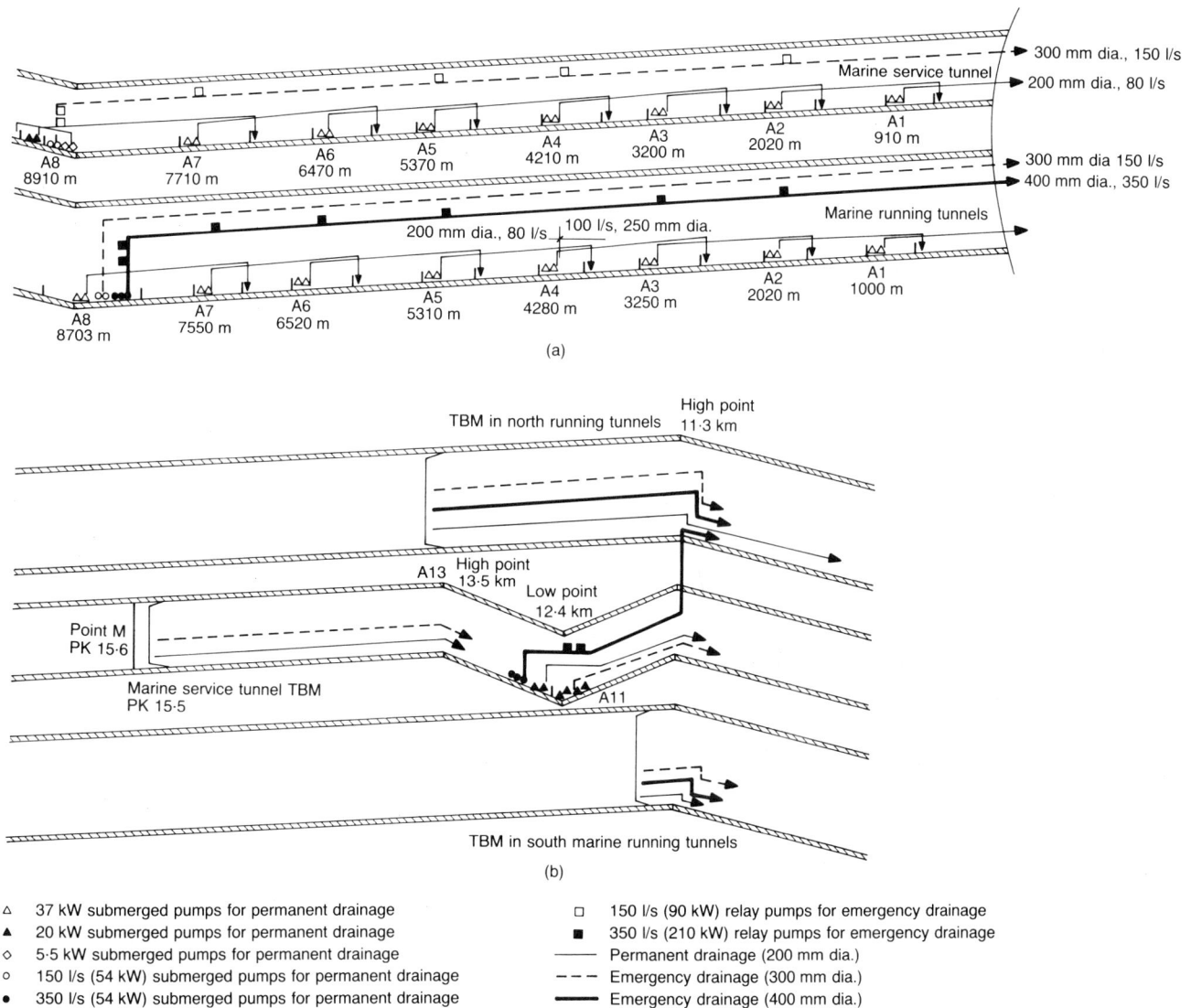

(a)

(b)

△ 37 kW submerged pumps for permanent drainage
▲ 20 kW submerged pumps for permanent drainage
◇ 5·5 kW submerged pumps for permanent drainage
○ 150 l/s (54 kW) submerged pumps for permanent drainage
● 350 l/s (54 kW) submerged pumps for permanent drainage

□ 150 l/s (90 kW) relay pumps for emergency drainage
■ 350 l/s (210 kW) relay pumps for emergency drainage
——— Permanent drainage (200 mm dia.)
– – – Emergency drainage (300 mm dia.)
▬▬▬ Emergency drainage (400 mm dia.)

Permanent drainage (service tunnel = 20 l/s, running tunnel = 25 l/s)

100 mm dia.

Waste trap
Air–water separator

6 m³ tank

Two pumps

Winding drum
60 m 100 mm dia.

200 mm dia.

200 mm dia.

Transfer pump

200 mm dia.
in tunnel

Flexible 100 mm dia.

Vacuum pump

Air supply

Air–water separator
250 mm dia.

Emergency flow (150 l/s)

250 mm dia.

Three winding drums
length 22 m 150 mm dia.

250 mm dia.

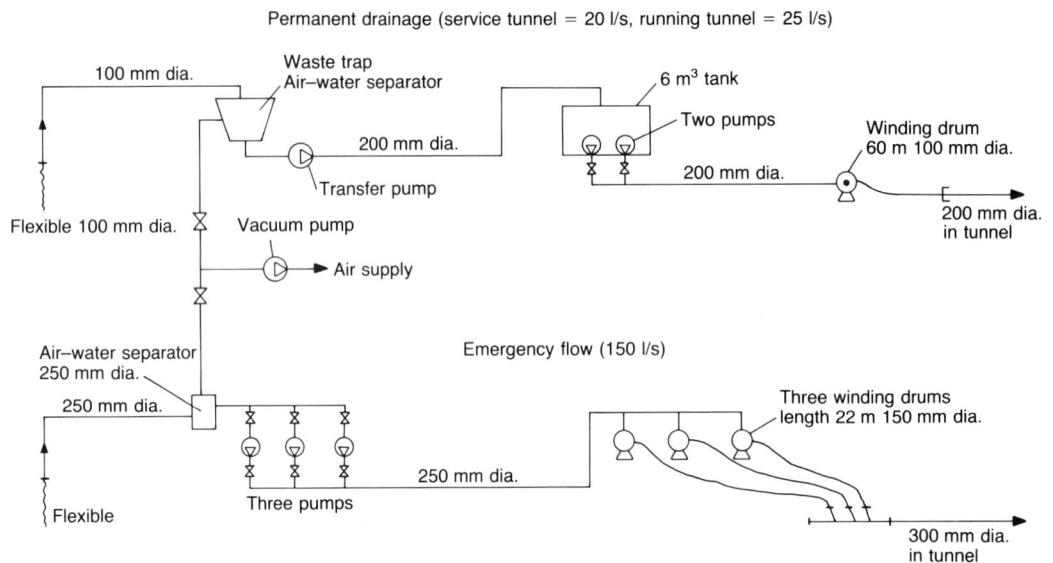

Fig. 2. Drainage at tunnel face on TBM back-up

Flexible

Three pumps

300 mm dia.
in tunnel

Drainage of the marine tunnels

13. The drive was downhill as far as the breakthrough of the service tunnel, except for a 2·5 km section with a slight uphill gradient (0·28%) beyond the first low point. Water therefore had to be raised by a head of 65 m between the low point and the shaft.

14. The four phases of drainage were as follows

(a) Phase 1—the period of tunnel driving before breakthrough of the service tunnel
(b) Phase 2—the period between breakthrough of the service tunnel and breakthrough of the running tunnels
(c) Phase 3—the period between breakthrough of the running tunnels and commissioning of pumping station W3
(d) Phase 4—the final phase.

Fig. 3. Tunnel dewatering pumps

Phase 1. Prior to breakthrough of the service tunnel (Fig. 1)

15. Each tunnel had two independent installations: a permanent, continuous system and an emergency system brought into service in case of inundation or when the permanent installation was being serviced or had broken down.

16. The drainage installation consisted of two distinct elements: a system at the face, fitted onto the TBM, which pumped water to a mobile pumping station sited about 2 km behind the TBM (Fig. 2) and a tunnel system, installed as the drive progressed, which returned water to the sump at the bottom of the shaft.

17. As seepage into the tunnels was significantly less than predicted at the outset, it was possible to base the size of the permanent network in the service tunnel on a 200 mm diameter pipe which provided for a maximum of 80 l/s, instead of the 98 l/s anticipated by the end of the tunnel drive.

18. By contrast, in the running tunnels 4 km from the shaft, the pipe had a diameter of 250 mm to cater for maximum discharges of 103 l/s. To ensure this rate of drainage on a descending slope, weirs were installed in the drainage channel in the tunnel invert at about 1 km intervals to create sumps. Two submerged pumps were installed in these sumps as well as at the low points of the tunnels (Fig. 3). The pumps discharged into a 200 mm diameter pipe which delivered into the next sump—and so on as far as the shaft.

19. Emergency drainage was provided by two networks in the running tunnels and one in the service tunnel. Each had a mobile pump station and relay pumps along the pipe at about 2 km intervals right through to the shaft. The service tunnel emergency network consisted of

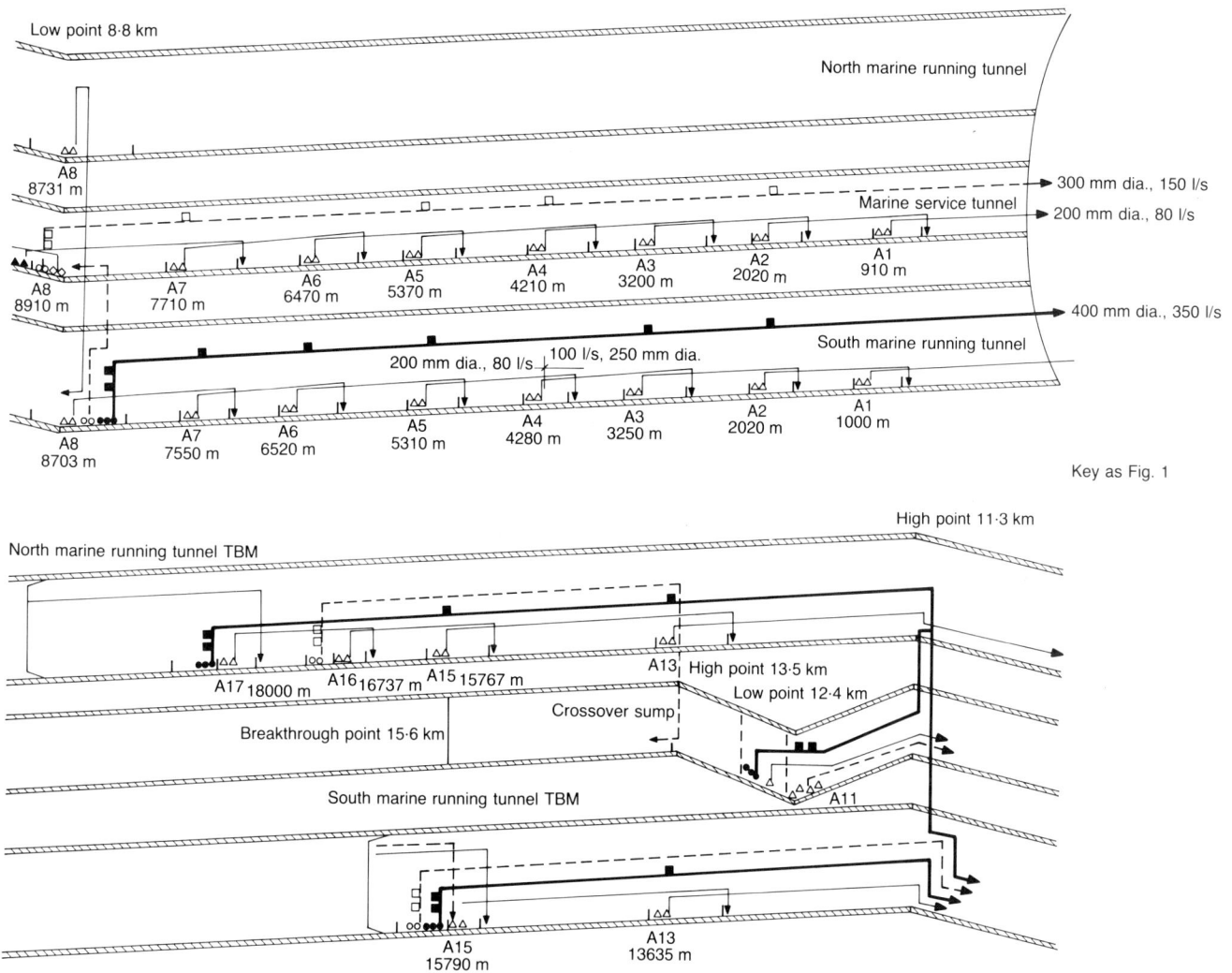

Fig. 4. Tunnel
drainage: Phase 2

a 300 mm diameter pipe with a capacity of
150 l/s; the network in each of the running
tunnels had a pipe of 400 mm diameter and a
capacity of 350 l/s.

*Phase 2. Breakthrough of service tunnel –
breakthrough of running tunnels* (Fig. 4)

20.   Following breakthrough of the service
tunnel, its permanent drainage system was
arranged for water to flow towards the UK from
the tunnel high point.

21.   Seepage from the lining between the two
high points of 2 l/s per km was collected in the
drainage sump in the crossover and pumped up
via the permanent system from the crossover
by three pumps of 40 l/s capacity and discharged through a 200 mm diameter pipe. This
water flowed by gravity to the low point of the
tunnel, and from there into the Phase 1 network.

22.   In the running tunnels, permanent
drainage between the face and the high point at
11·3 km from Sangatte shaft) did not differ
from that in Phase 1. It then relied on gravity to

flow to the low point where the permanent
drainage system served all the tunnels.

*Phase 3. Breakthrough of running tunnels –
commissioning of pumping station SW3*

23.   In the interval between the breakthrough of the north running tunnel and breakthrough of the south running tunnel, the
back-up drainage system of the latter (275 l/s)
was transferred to that of the north running
tunnel to flow by gravity towards the UK. The
same applied to the permanent drainage system
which was transferred to the service tunnel.

24.   The rate of permanent seepage was
assumed to be 1 l/s per km for a total of 34 km
of tunnels on the French side between the shaft
and the highest point in the tunnels. A temporary pumping station was used initially in
pumping station W3 north to provide drainage
at the rate of 34 l/s, using a submerged pump
sending water to the treatment plant via the
400 mm diameter permanent pipe in the service
tunnel.

*Proc. Instn
Civ. Engrs
Civ. Engng,
Channel Tunnel,
Part 3:
French Section,
1994, 30–33*

Paper 10494

Tunnels—ventilation

*H. Barthes, A. Bordas, D. Bouillot, M. Buzon, Ph. Dumont, J. Fermin,
J.-C. Landry, J.-P. Larive, L. Leblond, J.-J. Morlot, L. Szypura,
Ph. Vandebrouck and B. Vielliard*

■ **Ventilation requirements were evaluated
in accordance with the regulations and
recommendations of the social security
organization (the CNAM), the building and
public works safety organization, the
underground works association and the
underground mine regulations. This Paper
describes the equipment chosen for the
marine and land tunnelling operations and
explains how ventilation arrangements
changed as the work progressed.**

Various regulations and recommendations
(CNAM, OPPBTP, AFTES, Regulations for
Underground Mining) made it necessary to
define the following requirements for ventilation in the tunnels:

(a) diesel pollution: 50 l of fresh air per hp/s
(CNAM recommendation)
(b) at the face: 300 l/s of fresh air per m² of
cross-section of face (sites with dust
emissions)
(c) minimum fresh air per worker for the most
densely staffed shift: 25 l/s.

These figures were not cumulative; the highest
figure was always used.

Choice of ventilation system

2. It was decided to use electric traction for
the main trains, with power supplied by a combination of batteries and overhead lines. Only

manriders and emergency vehicles ran on
diesel.
3. The ventilation system had to supply
the three marine tunnels for distances from 0 to
20 km, with variations in progress between the
tunnels of up to 5 km. Studies showed that a
system of primary ventilation as far as the last
cross-passage and then supplementary ventilation via flexible pipelines was deemed a much
better solution than blowing fresh air through
ducts to the face all the way from the surface.
4. The system consisted of three phases

(a) start-up—air blown to the face from the
shaft by means of the tunnels' supplementary ventilation system and part of the
primary ventilation system at the surface
for extraction (Figs 1 and 2)
(b) mixed phase of both primary and supplementary ventilation—after the second
running tunnel had been driven beyond
2 km, the breakthrough of the first cross-
passages allowed primary ventilation to be
installed by extracting foul air in the north
running tunnel through an air lock and discharging it to the surface, the shaft and the
other two tunnels acting as air inlets; the
faces were ventilated by ducted blown air
from the most recently completed cross-
passages (Fig. 3)
(c) after final tunnel breakthrough, each
tunnel had an independent ventilation
plant about 16 km from the shaft equipped
with an air lock which had clearance for a
train to pass; the direction of air flow was
from France to the UK, with the ability to
reverse the system in an emergency such
as fire.

5. The relatively short length of the land
tunnels (3 km) gave rise to the selection of a
duct system of air blown from the surface to
the face. Foul air was handled in the marshalling chamber and was extracted via the shaft
and discharged at the surface by a fan. After
breakthrough, the natural ventilation flow in
the land tunnels was from the shaft to the
Beussingue Portal. This was at the rate of
30 m³/s in the running tunnels and 10 to
15 m³/s in the service tunnel, both in winter
and summer.
6. In view of the choice of electrically
driven trains, the ruling parameter for ventilation in the tunnels was the flow of 300 l/s of
air per m² of exposed face to and from the face,
which called for

*Fig. 1. Flexible
overhead blower pipe
(Ventube) taking
fresh air from
marshalling chamber
to tunnel face: top
left is start of
extraction duct*

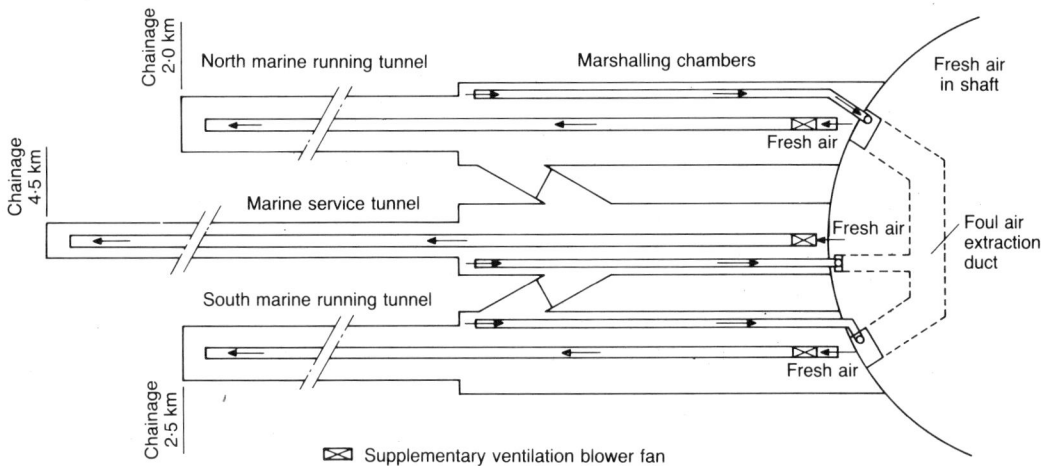

Fig. 2. Marine tunnels ventilation: start-up phase

Supplementary ventilation blower fan

(a) 5·4 m³/s of fresh air in the service tunnel (internal diameter 4·8 m)
(b) 13·5 m³/s of fresh air in the running tunnels (internal diameter 7·6 m).

In normal operation, the shaft was used solely as an air inlet to avoid the formation of fog arising from the discharge of warm foul air at the base of the shaft.

Marine-side ventilation system

7. The full cross-sections of the south running tunnel and the service tunnel were used to supply fresh air from the shaft. The full cross-section of the north running tunnel was used to allow stale air to be returned through a duct passing above an air lock (which insulated the tunnel from the marshalling chamber) and extending up the side of the shaft to the extraction plant at the surface. Discharge was therefore outside the shaft.

8. Primary ventilation was provided up to where the last cross-passages and pressure relief ducts had been opened up. The working faces were ventilated by blown air in flexible tubes to the end of the back-up (Fig. 4). The north running tunnel was supplied from the south running tunnel through an opened piston relief duct.

9. Foul air was returned via the north running tunnel from the south running tunnel via the last piston relief duct and from the service tunnel via the last cross-passage. The preceding ducts were either all closed or were fitted with air locks.

10. Ventilation at the face was provided by a ventilation duct on the back-up, which returned foul air through a dust filter behind the outlet of the fresh air supply duct (Fig. 5).

11. On breakthrough of a tunnel, ventilation equipment fitted with an air lock was installed at the breakthrough point and the supplementary ventilation was removed.

12. The size of the surface level fan which provided primary ventilation was determined by the maximum length needed. In the service tunnel this was to chainage PK 18 km (2 km beyond the theoretical breakthrough point), and in the running tunnels it was to chainage PK 13·5 km. The surface level fan therefore had a discharge rate of 180 m³/s and an operating pressure of 450 mm of water (useful pressure 385 mm of water).

13. It had to be possible to ensure supplementary ventilation both of the service tunnel for a maximum length of 7·5 km (maximum advance over the running tunnels), and of the running tunnels for a maximum length of 4·5 km (delay in opening up cross-passages). The characteristics of supplementary ventilation for each running tunnel were

(a) Ventube blower pipe: diameter 1600 mm, length 4500 m
(b) one fan per tunnel: discharge rate 22 m³/s, pressure 290 mm of water.

Fig. 3. Marine tunnels ventilation: primary and supplementary systems

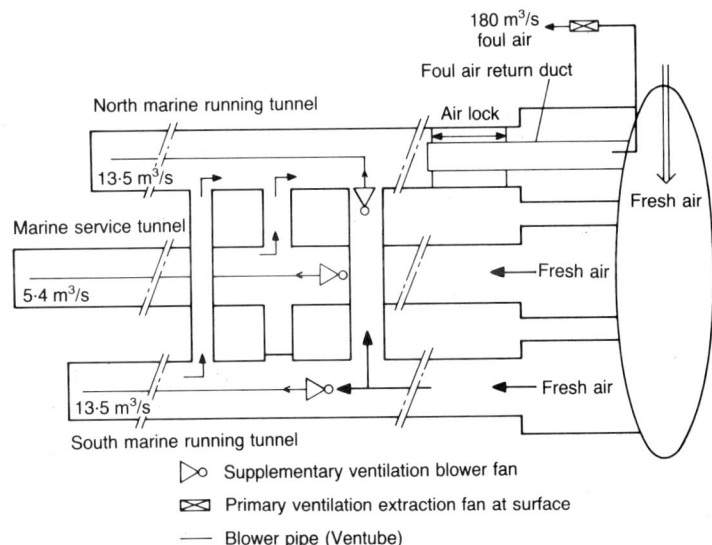

Supplementary ventilation blower fan

Primary ventilation extraction fan at surface

Blower pipe (Ventube)

Fig. 4. Blower pipe arriving at TBM

Characteristics of supplementary ventilation for the service tunnel were

(a) Ventube
(i) diameter 1200 mm, length 6700 m,
(ii) diameter 900 mm, length 830 m up to the level of the grouting gantries;
(b) three fans in line: discharge rate 18 m^3/s, pressure 380 mm of water.

14. Discharge from supplementary fans was increased by 10% to improve the temperature conditions during the start-up phase of boring. Allowances for leakage were in keeping with the high quality of these installations.

Land-side ventilation system

15. The land service tunnel and south land running tunnel, each 3000 m long, were ventilated by a blown air system in flexible tubing. The fan was installed at the entrance to the marshalling chamber and air intake was from the shaft. The air left the system at the level of the back-up. An extraction pipe fitted with a dust filter collected the air at the face and discharged it behind the blower pipe. The foul air was handled in the marshalling chamber via an extraction pipe with a fan at the surface.

16. The north land running tunnel was similarly equipped, but started from the Beussingue Portal. After breakthrough, the air extraction system was retained and converted to provide a ventilation system from the Portal to the shaft in case of fire.

17. In terms of scale, the principles were the same as for the marine tunnels

(a) one fan at the surface discharging 37 m^3/s of air at a pressure of 105 mm of water
(b) in the service tunnel
(i) a 3000 m long, 1200 mm diameter Ventube
(ii) an extraction fan operating at 8 m^3/s at a pressure of 170 mm of water
(c) for each of the running tunnels
(i) a 3000 m long, 1600 mm diameter Ventube
(ii) a fan extracting at 18 m^3/s and a pressure of 190 mm of water.

Other aspects affecting the choice of ventilation system

18. The ventilation system was sized to ensure sufficient air flow at the face, but it also had to satisfy the criteria for extracting gases (hydrogen and carbon dioxide), for the speed of air flow and the resultant environmental conditions.

19. Calculations on eight trains in the service tunnel and nine in each of the running tunnels when recharging batteries indicated that at least 21 m^3/s of air was necessary for all three tunnels to dilute the resultant hydrogen to 0·8%. In the service tunnel at the face, if all the hydrogen were concentrated there, it would need 2·8 m^3/s ventilation but 5·4 m^3/s was provided; and in the running tunnels at the face, 9·5 m^3/s would be needed but 13·5 m^3/s was provided. The ventilation system was, therefore, more than adequate.

20. In the marine section it was envisaged that a maximum of 150 diesel hp would be in use at any one time in the three tunnels, i.e. the need was for 7·5 m^3/s of air. This requirement was assured because 32·4 m^3/s was available (5·4 + 2 × 13·5). In the land section 50 hp would be used, i.e. a minimum of 2·5 m^3/s was needed but at least 5·4 m^3/s was available.

21. Return air cycle velocity was more than 4 m/s by the air lock through which air returned to the north running tunnel. However, at the piston relief duct which returned air from the south running tunnel to the north running

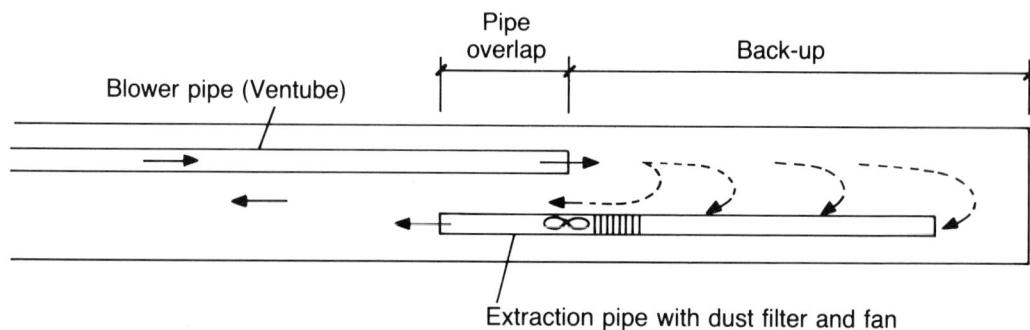

Fig. 5. Ventilation at tunnel face

tunnel, the speed could reach 10 m/s locally. There was no intention that personnel would be in this duct.

22. The principal aim was to ensure that maximum temperature would be in the region of 28°C. The calculation was based on the heat emissions from the cutting head (8 kWh/m³ of excavated chalk), the equipment and motors of the TBMs, the locos and transformers installed in the passages. The assumed daily rate of drive was 30 m per tunnel. All the sources of heat were partly balanced by that which could be absorbed by the cold seepage water and the surrounding ground. Surplus heat was extracted by the ventilation system. The further the TBMs progressed, the more significant became the heat absorbed by the surroundings, while the heat source varied little. The most extreme temperatures therefore arose during the early part of tunnelling. Calculations for such conditions at their worst—dry ground, a temperature of 12°C and dry excavation— showed that increasing extraction to the limits of the capacity of the supplementary systems would keep the highest temperatures lower than 28°C.

*Proc. Instn
Civ. Engrs
Civ. Engng,
Channel Tunnel,
Part 3:
French Section,
1994, 34–38*

Paper 10495

Tunnels—shaft and marshalling chambers

*H. Barthes, A. Bordas, D. Bouillot, M. Buzon, Ph. Dumont, J. Fermin,
J.-C. Landry, J.-P. Larive, L. Leblond, J.-J. Morlot, L. Szypura,
Ph. Vandebrouck and B. Vielliard*

■ **The shaft was a temporary structure used
for the driving of all tunnels from a single
intermediate working point; after these
works, it was used to house the permanent
ventilation and pumping equipment. This
Paper describes the shaft and its construction. Marshalling chambers were enlarged
sections of tunnels allowing the assembly
of TBMs and back-ups and then the
manoeuvring of service trains during
tunnel drives. The Paper describes their
construction method, which had to be
adapted according to the terrain encountered.**

Functions

The shaft was a temporary structure which
allowed all tunnels to be driven from a single
intermediate heading. Its principal functions
were to

(a) provide access to the level of the tunnels
    while being protected from the water-table
(b) allow the TBMs and back-ups to be
    lowered in sections prior to their re-
    assembly in the marshalling chambers
(c) provide access and provision of supplies to
    the tunnelling sites
(d) allow removal of spoil by using the base of
    the shaft to house the pumping station con-
    nected to the spoil disposal site
(e) allow extraction of seepage water.

2. The marshalling chambers were enlarged
sections of the first 50 m of the tunnels on each
side of the shaft. They permitted initial
assembly of the TBMs and back-ups, and movement of supply trains and removal of spoil
while the tunnels were being driven.

3. On completion of the works, the shaft
was reorganized to receive service ventilation
and permanent pumping equipment.

Description

4. The shaft was constructed of reinforced
concrete, was cylindrical with a flat base, had a
55 m internal diameter and was 65 m deep.

5. The six marshalling chambers were
created on opposite sides of the shaft, each
50 m long and 48 m below ground level. The
service tunnel was enlarged to 8·9 m wide, and
the running tunnels to 11 m for the purpose.
Cross-passages connected the running tunnels
to the service tunnel at the level of the marshalling chambers.

6. The shaft and marshalling chambers
were surrounded by a 60 m deep cut-off wall of
bentonite cement excavated by jetting to the
level of impermeable chalk to give protection
from the water-table while the tunnels were
being driven. In plan, the wall is in the form of
a 200 m by 100 m ellipse, the tunnels running
along the long axis (Fig. 1).

Construction of the shaft

7. The shaft was sunk from ground level
(+18·30 masl) through the sandy and chalky
silts of the Quaternary period to about
−23 masl, then through the White and Grey
Chalks to its base at −46 masl. The water-table
was at +2 masl.

8. From level +18·30 m (i.e. +18·30 masl)
to +6·30 m, excavation was inside a circular
diaphragm wall by means of three back-actor
shovels discharging into lorries which removed
the spoil using a peripheral ramp (Fig. 2).

9. From level +6·3 m to the base of the diaphragm wall at level −3 m, then to level
−25·25 m, excavation was by means of three
back-actor shovels discharging into skips of
10 m³ capacity raised by crane (Fig. 3). The
chalk was loosened by a D10 fitted with a
single-toothed ripper.

10. The shaft lining of successive rings of
2·5 m was poured in three lifts in each of four
sections of 14·4 m of arc. Each lift involved successive stages: excavation, reinforcement, shuttering and concreting. The work proceeded at
most on two levels. Four days were required to
complete a ring.

11. The reinforcement cages were made up
of two layers. Horizontal linkage was by means
of starter bars extending from the preceding
section; the vertical rods were linked by
straight bars with sleeves. The all-metal formwork was hung from ballasted carriers which
ran along the coping beam. Concreting was
carried out with two 180 t/m tower cranes
handling 1 m³ capacity side-tipping skips.

12. Working was in three eight-hour shifts,
seven days a week. Reinforcement, formwork
and concreting were completed successively in
one shift (Fig. 4).

13. Work in the shaft stopped at level
−25·25 m so that the marshalling chambers

Section A–A

Plan

could be constructed. This was done after the surrounding trench had been dug in two stages—crown headings from level −26 m, and sidewall drift and inverts from level −30·25 m.

14. Between levels −20·25 m and −30·25 m, formers were placed in the lining to make reservations for the openings to the marshalling chambers. The lower section of the shaft between levels −30·25 m and −46 m was constructed in the same way as the upper section. Two excavators and a D10 were used in the base of the shaft with two tracked cranes at the surface to handle the 10 m³ capacity skips. Five lining rings identical to those above −25 m were placed (Fig. 5). A reinforced concrete raft foundation 1 m thick was laid below the last ring at level −46·25 m.

15. Four 2 m diameter piles to support the steel decking at level −27 m were driven from

level −30·25 m to level −51 m by traditional methods, reinforced and concreted throughout their height (Fig. 6).

Construction of the marshalling chambers

16. *Crown headings* The upper half of the service tunnel, the entries to the running tunnels and the headings for starting the TBMs were excavated in successive 3 m passes using a Westfalia-Lynx road header of 100 kW capacity. The spoil was conveyed to the shaft by front-end loaders of 3 m³ capacity.

17. Ground stabilization was by shotcrete onto mesh and 3 m long rock-bolts bonded with resin. The concrete was sprayed by the wet method with a lance mounted on a robot arm.

18. The top headings of the running tunnels beyond the tunnel openings were driven by a

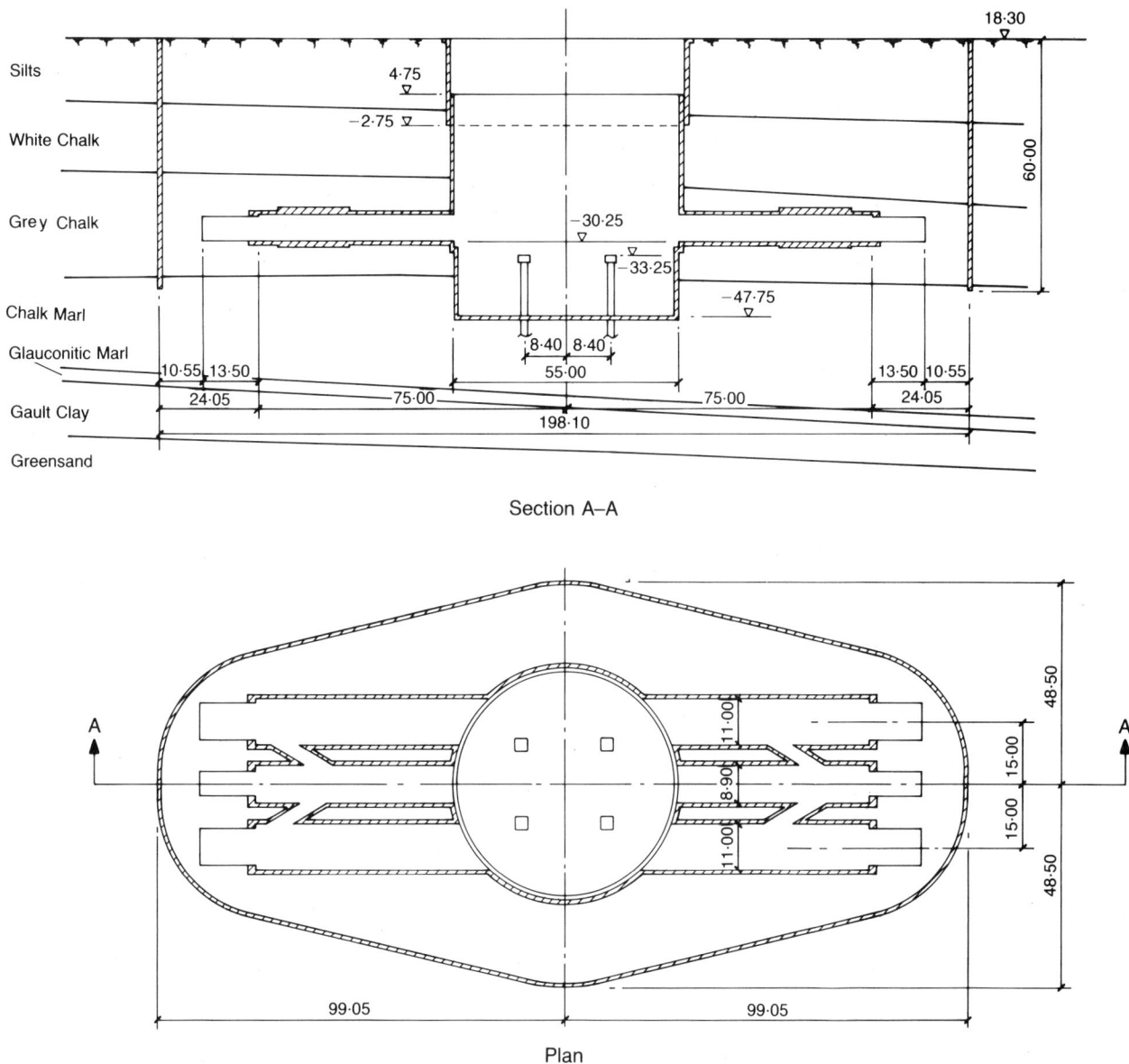

Fig. 1. General arrangement of shaft and marshalling chambers (dimensions in m, elevations in masl)

Fig. 2. First phase of
shaft construction
inside circular
diaphragm wall

Fig. 3. Second phase
of shaft construction
with lining cast
in situ

different method. Given the geological conditions met when excavating the marshalling chamber of the service tunnel (heavily fractured chalk), the excavation method was replaced by a technique of pre-cutting the chalk by mechanical means which involved sawing ahead into the ground. A cut was made corresponding to the curve of the vault, injected with concrete, and the tunnel was excavated under the protection of this preliminary vault, further supported by two metal arch beams.

19.   In the service tunnels, the traditional method of constructing the roof lining in 6 m sections simultaneously a short distance behind the excavation was employed. The shuttering and concreting cycle was carried out in three shifts, using special metal shuttering. Since the rate of excavation was 3 m a day, the gangs worked in each tunnel in turn while the concrete was setting.

20.   In the running tunnels, the amount of space taken up by the pre-cutting machine and the lack of space to stock materials and move about on the access deck meant that excavation and roof lining could not proceed simultaneously. Lining the arch was therefore carried out by the method applied in the service tunnel but in 12 m lengths using two sets of formwork.

21.   Construction of the lower half of the marshalling chambers was carried out in two stages: the invert after excavating the core

		1987												1988					
		J	F	M	A	M	J	J	A	S	O	N	D	J	F	M	A	M	J

Shaft
Elliptical cut-off wall
Circular diaphragm wall
Shaft to −23 masl
Lining around openings and piles
Shaft base to −47 m

Marshalling chambers
Entrance to chambers
Upper half (crown,   ST(m)
tympanum, entry)   RTN(m)
     RTS(m)
     ST(l)
     RTN(l)
     RTS(l)

Lower half (crown,   ST(m)
tympanum entry)   RTN(m)
     RTS(m)
     ST(l)
     RTN(l)
     RTS(l)

Cross-passages
Sidewall, drifts
Crown, invert

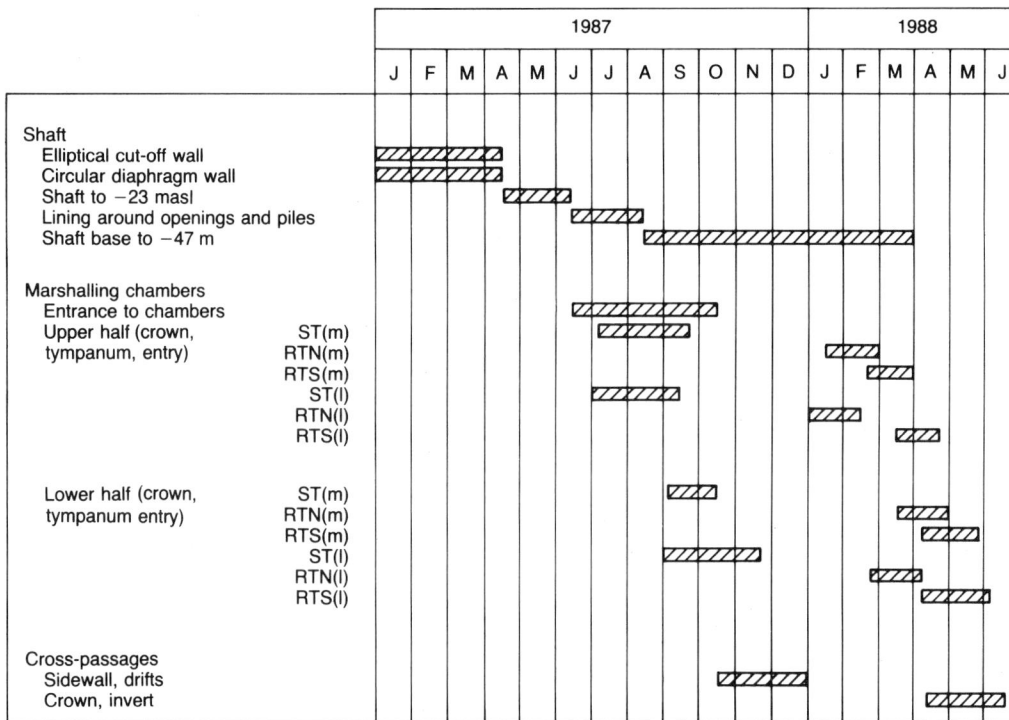

Fig. 4. Shaft construction schedule

Fig. 5. Shaft base construction below openings for marshalling chambers

37

Fig. 6. Shaft base showing steel decking support piles

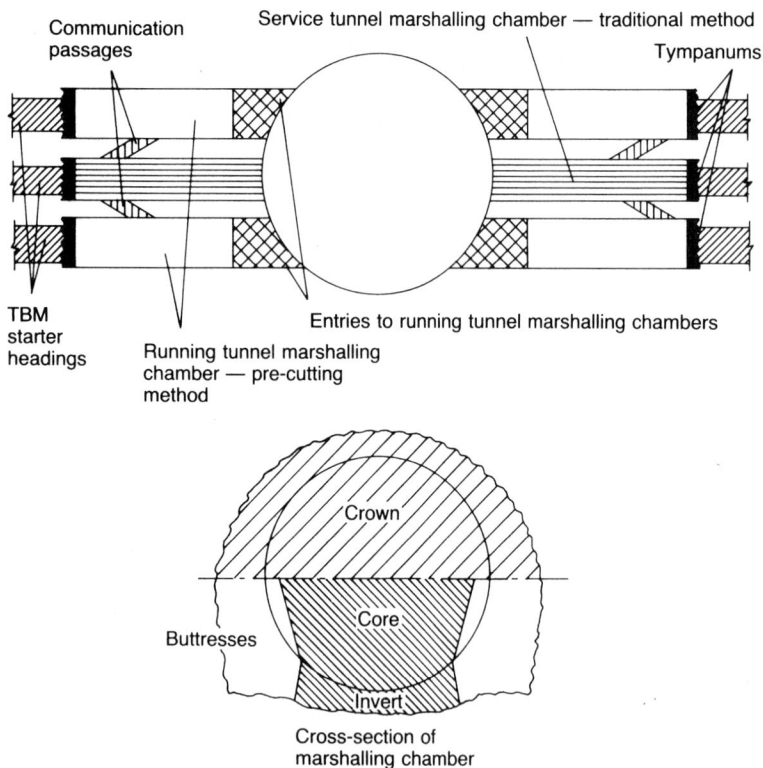

Fig. 7. Marshalling chamber construction methods

leaving an abutment below the springing of the vault, and the buttresses, in the return direction, from the end in alternate passes (Fig. 7).

22. The core was dug by a D10 with a ripper. Spoil was extracted and delivered to skips of 10 m³ capacity alongside the shaft. The invert was constructed by 6 m pours between two guides laid down during the advance. The buttresses were dug by hydraulic shovel in alternate sections 3 m long. Face stability was achieved by rock-bolting and shuttering to the sections was supported by resting on the fixings at the base of the arch and along the invert.

23. The buttresses in the marine and land tunnels were poured simultaneously, the gangs working in each alternately while the concrete set. The excavation, shuttering, concreting and setting sequence was completed in four shifts.

24. The tympanum and the tunnel opening where the TBMs started were constructed in the same way as the main tunnels.

25. The tympanums were shuttered and concreted, the openings excavated by road header and lined in shotcrete sprayed on metal arch gauges. Face stability was achieved by rock-bolting.

Tunnels—the equipment of the shaft and marshalling chambers

H. Barthes, A. Bordas, D. Bouillot, M. Buzon, Ph. Dumont, J. Fermin, J.-C. Landry, J.-P. Larive, L. Leblond, J.-J. Morlot, L. Szypura, Ph. Vandebrouck and B. Vielliard

*Proc. Instn
Civ. Engrs
Civ. Engng,
Channel Tunnel,
Part 3:
French Section,
1994, 39–43*

Paper 10496

■ **As the shaft was the only means of access to the working faces of the tunnels, its equipment was of great importance. This Paper reviews the general design of equipment of the shaft and marshalling chambers and distinguishes the works carried out on the surface from those down below. The equipment required for the complex organization of train traffic in the marshalling chambers is dealt with in detail.**

Organization

Only those items of equipment indispensable to the functioning of the works or which were necessary to keep up the rate of work were installed in the shaft

(a) access from the surface to the level of the tunnels (two stairways, three hoists, two lifts)
(b) three automatic systems for supplying segments
(c) five stations at the base of the shaft for tipping spoil from the wagons
(d) five crushers and slurry mixers; the slurry was then pumped by eight sets of piston pumps from the bottom of the shaft and conveyed to the Fond-Pignon discharge site
(e) drainage pumps and sumps in the base of the shaft
(f) areas for stocking and loading concreting and grouting materials
(g) a recharging point for locomotive batteries
(h) a central control post controlling all the shaft installations and the tunnel face works.

Equipment at the head of the shaft

2. Above ground, the equipment and installations needed to service the segment requirements led to the installation of two beams crossing the shaft to support the following structures

(a) a central covered nave running north to south for supplying segments to the running tunnels, equipped with two automated travelling cranes of 60 t capacity. This supported the handling framework for the TBMs, capable of transferring and lowering a load of 430 t for assembling the TBMs (Fig. 1).

Fig. 1. 430 t gantry crane at the top of the Sangatte shaft

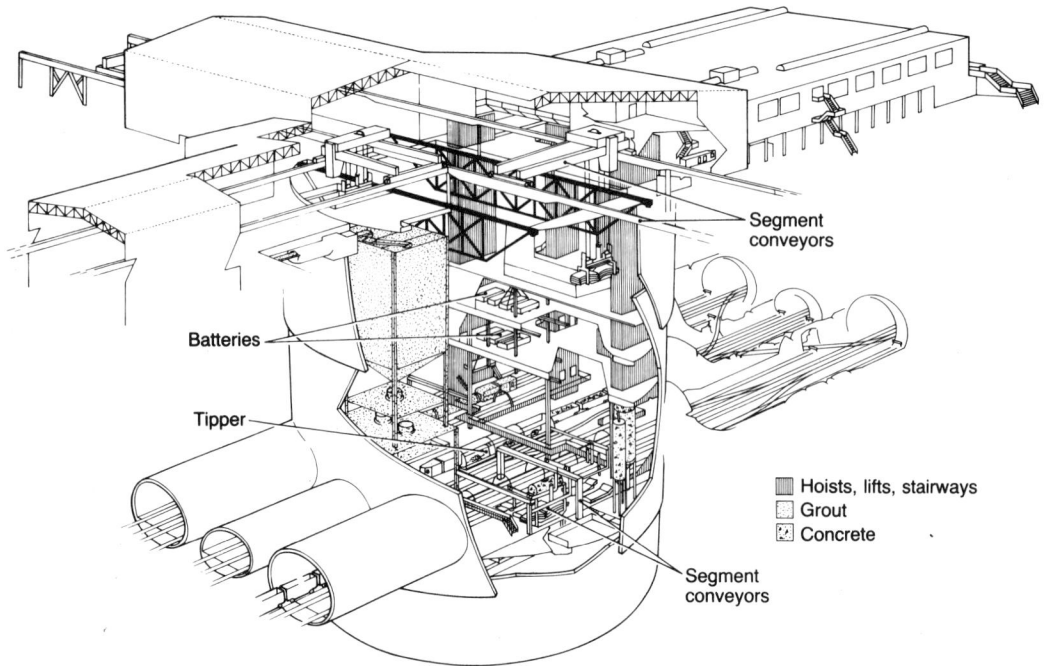

Fig. 2. The Sangatte shaft and marshalling chambers: overall cutaway view

(b) a second nave (east) adjacent to the first, on the land side, dedicated to the supply of segments for the service tunnel by an automated travelling crane of 30 t capacity

(c) on the side nearest the sea was an assembly point for the transfer of staff, small items of equipment and maintenance between the surface and the base of the shaft. Below this was the administrative centre of the technical services for equipment and installation in the shaft and tunnels, and the control post.

See Figs 2 and 3.

3. The supply of segments involved the following operations

(a) stocking prefabricated segments (before palletization) sufficient for two months' TBM drive

(b) grouping complete rings on special palettes and moving them by motorized carrier to the palette storage area

(c) stocking about 150 palletized rings including the special rings for the cross-passages and a fall-back stock of standard rings

(d) lowering of stock into the shaft by special handling equipment for each of the three

Fig. 3. Aerial view of the Sangatte site showing the covered naves over the shaft (centre)

tunnels (service tunnel, north running tunnel and south running tunnel)

(e) two surface traverse conveyors each carrying ten rings to cater for peak demands

(f) a 60 t travelling crane for each running tunnel and a 30 t crane for the service tunnel

(g) two reception traverse conveyors at the base.

Reception equipment at the base of the shaft

4. Equipment for access to the base of the shaft consisted of personnel lifts and dual-purpose hoists between the loading area at level +18 masl and the bottom reception area at level −28 masl

(a) twin lifts in one shaft, each 1800 kg or 12 persons capacity, for the use of personnel involved in the shaft installations

(b) three dual-purpose hoists of 7·5 t capacity, either 80 persons standing, or one manrider (for personnel, fire-fighting, etc.) for the personnel in the tunnels.

5. Battery charging and maintenance rooms were located in the shaft at two levels, −7·5 masl and −15 masl, there being two rooms each of 575 m² floor area. Handling was performed by a monorail or a double rail on each level. Eight battery recharge points of 450 Ah were provided for locomotives and tractors to the service tunnels, and 16 of 700 Ah for locomotives to the running tunnels. A special ventilation system with extraction at ceiling level and gas detectors was installed to minimize risk.

6. The reception area at level −28·2 masl provided the following equipment

(a) five tippers (one for each tunnel) by means of which the spoil wagons of each train could be tipped in two cycles, three wagons at a time for the service tunnel tipper and six at a time for each of the two tippers of the running tunnels (Fig. 4)

(b) two traverse conveyors for pallets of segments for each of the three tunnels

(c) refill equipment for tankers carrying concrete and grout

(d) a rail track network allowing movement of trains in the station. A train arrived every 20 minutes of each drive (i.e. for two simultaneous drives in a marine and a land tunnel, tipping, segment and grout supply services catered for a ring every ten minutes). In addition, the cross-passages in the marshalling chambers between the running and service tunnels, both land and marine, allowed full trains from the marine tunnels to cross the shaft to emerge at the Beussingue Portal after breakthrough of the land service tunnel in March 1989.

Fig. 4. Spoil wagon tipper in the reception area of the Sangatte shaft

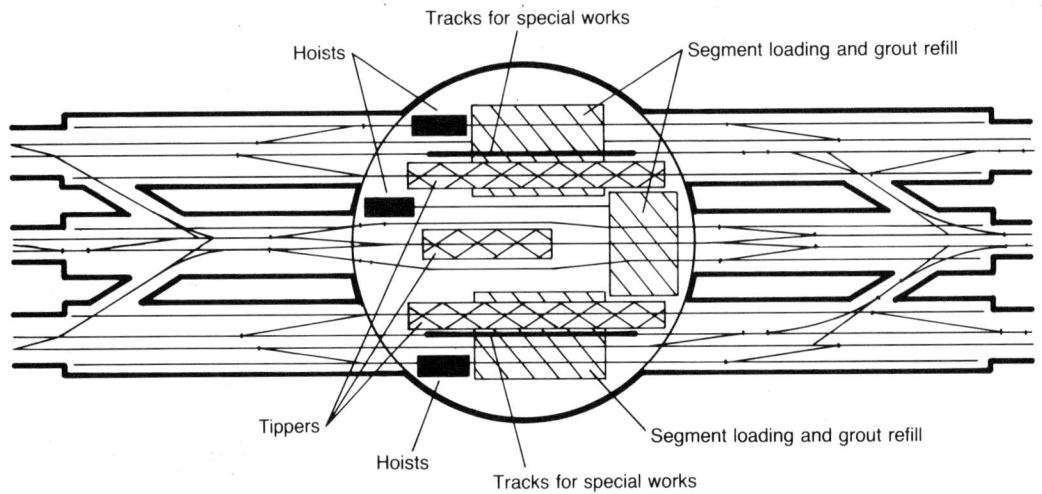

Fig. 5. Rail track system in the Sangatte shaft and marshalling chambers

Fig. 6. Slurry mixer at the base of the Sangatte shaft

Rail track system

7. The rail track system was complex; it was designed to meet the following requirements (see Fig. 5)

(a) tipping of spoil wagons
(b) supply of segments and grouting materials
(c) stockpiling of pallet cars
(d) crossover for trains serving TBMs
(e) siding for trains not in use
(f) arrival at the lower reception area, via hoist, of manriders and special vehicles (stored at the surface)
(g) assembly and equipment of trains for the special works.

8. Each marine and land tunnel was organized independently of the others, although connections through the cross-passages at the stations were set up between them for temporary installation activities or breakdowns, transfer of rolling stock between tunnels, and for traffic to the terminal through the service tunnel as soon as the latter had been driven through. This area had more than 80 sets of points.

Spoil handling plant

9. The pumping facilities at the base of the shaft comprised, from top to bottom, starting

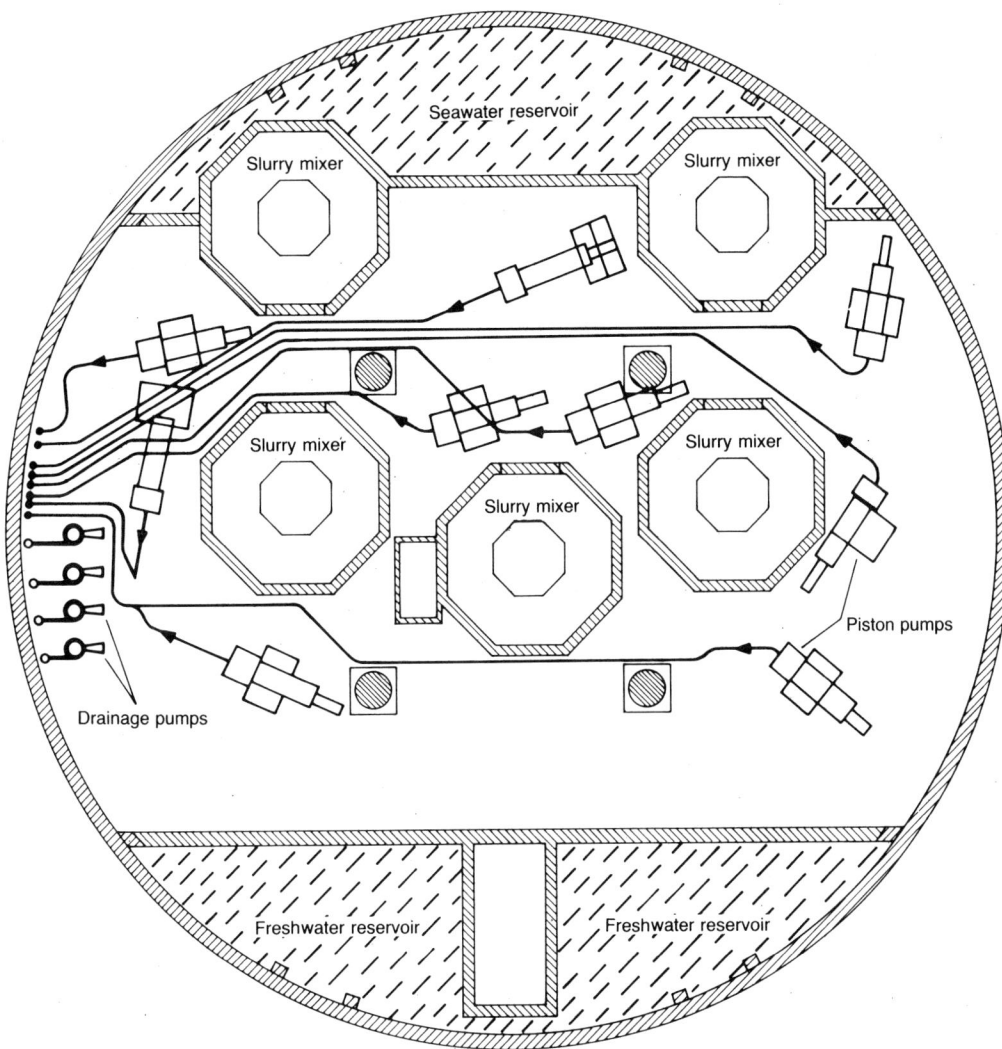

Seawater reservoir

Slurry mixer

Slurry mixer

Slurry mixer

Slurry mixer

Slurry mixer

Piston pumps

Drainage pumps

Freshwater reservoir

Freshwater reservoir

Fig. 7. Spoil pumping station at the base of the Sangatte shaft

from the level of the tippers at level −28 masl

(a) five feeder hoppers situated over three conveyor feed lines 2 m wide, i.e. five feeders—two for each running tunnel, one for the service tunnel

(b) five metal chutes each channelling the spoil to its own twin crusher fitted with toothed rollers

(c) The crushers fed five slurry mixers each consisting of a tank in the form of an octagonal funnel in which four arms fitted with harrows turned (Fig. 6). By manually regulating the water supply a chalk slurry was produced which could be pumped away

(d) this slurry flowed naturally through a 10–15 mm mesh grid onto a concrete floor at level −43·8 masl above the extraction pumps fed by way of traps

(e) at the base of the shaft, at level −47·75 masl, eight Putzmeister KOS 2180 piston pumps fed eight 250 mm diameter pipes delivering straight to the Fond-Pignon discharge site (Fig. 7).

*Proc. Instn
Civ. Engrs
Civ. Engng,
Channel Tunnel,
Part 3:
French Section,
1994, 44–47*

Paper 10497

Tunnels—driving

*H. Barthes, A. Bordas, D. Bouillot, M. Buzon, Ph. Dumont, J. Fermin,
J.-C. Landry, J.-P. Larive, L. Leblond, J.-J. Morlot, L. Szypura,
Ph. Vandebrouck and B. Vielliard*

■ **The Paper makes a clear distinction between the two driving methods used in the tunnelling process: the open mode and the closed mode, the first being theoretically faster but slowed in practice by the spoil handling and disposal capacity. The tunnelling machines, of complex design, mobilized a large amount of computer-controlled and automatic equipment, also described in this Paper.**

The two methods

The tunnel boring machine (TBM) excavated the chalk by the simultaneous process of rotating the cutter head and exerting forward thrust against the working face. Rotation was achieved by electric motors which operated a crown wheel integral to the cutter head; forward thrust was exerted by rams.

2. The thrust rams and the rear gripper unit thrusted temporarily against the chalk face or the lining. The rear gripper rams ensured a constant minimum thrust against the lining to compress the waterproof joints between the rings.

Open mode

3. When the TBM was working in open mode, the rear of the shield was kept in position by the grippers in contact with the rock. The front thrust rams reacted against it and pushed the telescopic cutter head forward beyond the shield (in up to 1·4 m increments on the service TBM and 1·6 m on the running tunnel TBMs) (Figs 1 and 2). Pressure was applied to the segments at the same time.

4. At the end of the cycle, the gripper rams retracted and released the rear shield. The rear shove rams thrust against the last ring placed and the shield moved forward (Fig. 3). The TBM thus moved like a caterpillar.

Closed mode

5. In closed mode the cutter head remained retracted (Fig. 4). When cutting, both front and rear shields were moved forward by the rear shove rams exerting pressure on the lining (Fig. 5). At the end of the drive cycle these rams were retracted, thus creating space in the rear to place the next segments (Fig. 6).

6. In theory, the open mode is a quicker method because advancement and segment placing occurs simultaneously. In practice, logistics were a constraint and the advantage of driving in open mode was not great. The capacity to treat and remove spoil in the shaft in effect limited the rate of advance.

Computerized control of the TBMs

7. In a machine as complex as the TBM, automation and computerization played leading roles. Some 1300 items of data were needed to activate the 700 motors, switches, rams, pumps and motorized valves.

8. The operator was in contact with other workers on the TBM and on the supporting train by radio and telephone. The control panel facing the operator consisted of dozens of switches and illuminated dials which enabled him to perform the operations and checks outlined below (Fig. 7).

9. He controlled the operation of pumps and motors, selected the drive method, started the screw and the conveyors, decided on the direction of rotation and the speed of the cutter head and operated the rams, grippers, telescopic head etc.

10. He adjusted the angle of the machine, based on information received from the ZED guidance system and the clinometers and, in accordance with instructions on the ring data sheet

(*a*) monitored, on some 50 numerical and 100 visual displays, the activity of the principal elements

  (i) for the cutter head disc: speed of the twelve motors, power and torque, rate of water injection, pressure on the front bulkhead in five different places
  (ii) for the front and rear shove rams: extended length, speed and hydraulic pressure
  (iii) for the gripper units: hydraulic pressure
  (iv) for each of the two screws: rotating speed, torque, hydraulic pressure, and pressure of the muck on entering and leaving
  (v) for the tail joint: mastic pressure

(*b*) he took readings, from a large screen, of real time information from the tunnelling supervisor

(*c*) he kept a check, by way of four monitors, on the trains (by means of two cameras placed at each end of the back-up), the extraction screw and the spoil conveyor (towards which three other cameras were

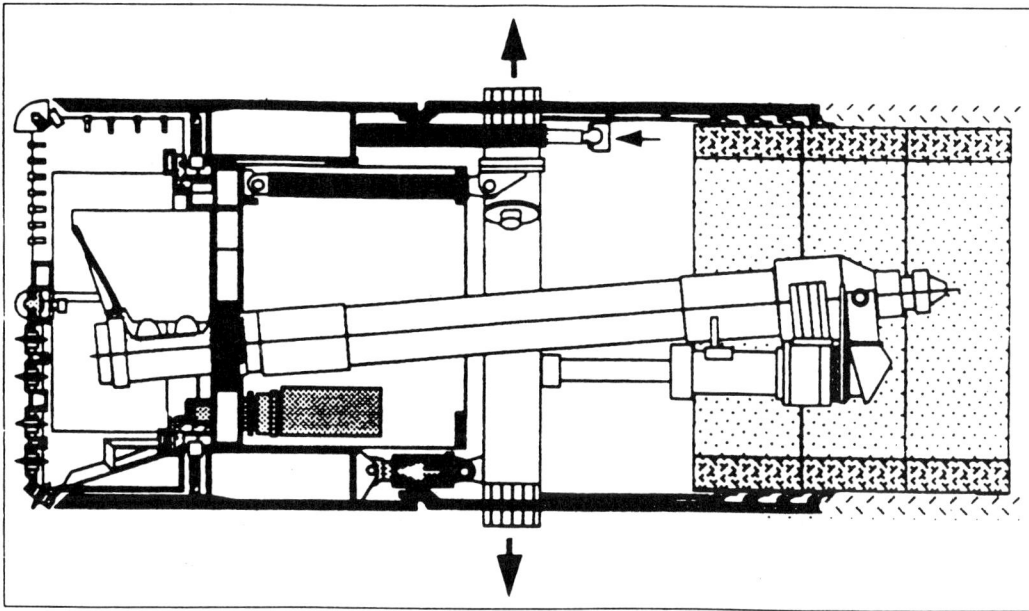

Fig. 1. Open mode, phase 1: extension of grippers; steering adjustments retraction of rear thrusters

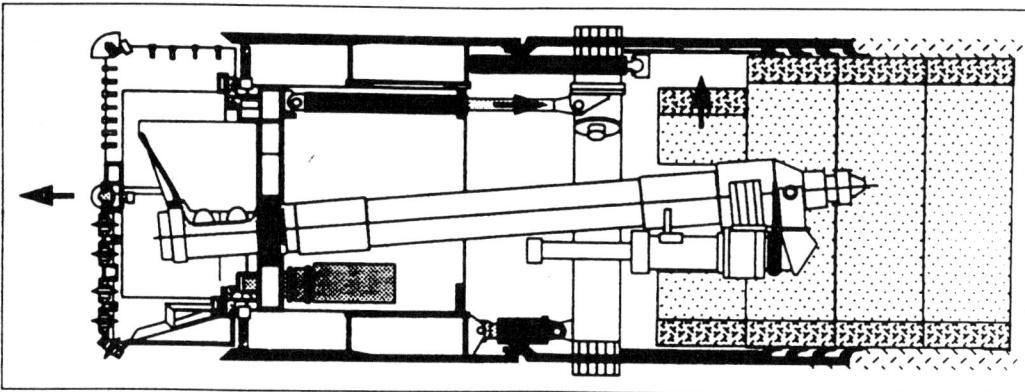

Fig. 2. Open mode, phase 2: start of boring; advance of telescoping head under front thrusters; simultaneous placement of segments

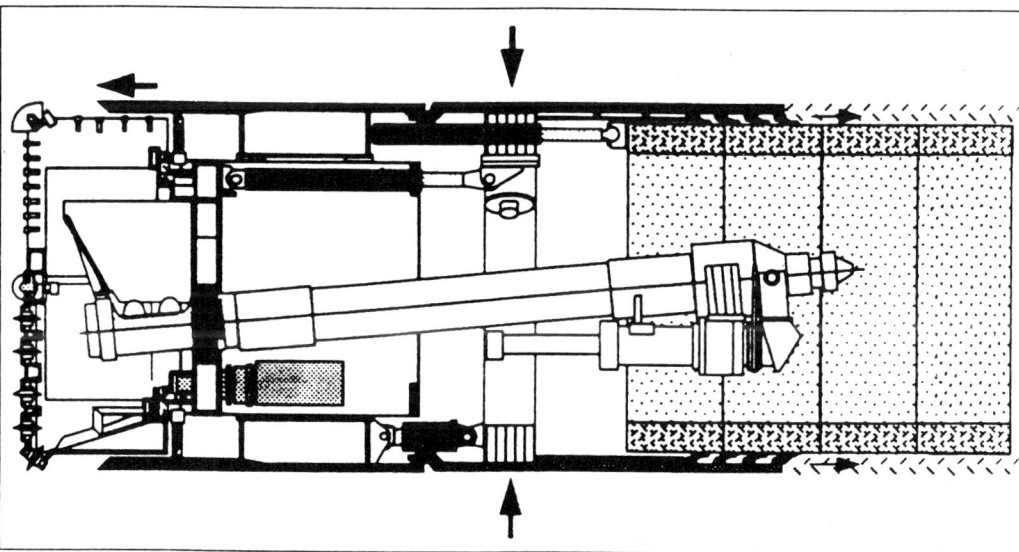

Fig. 3. Open mode, phase 3: retraction of grippers; application of rear thrusters on segments; retraction of front thrusters and advance of shield covering the cutting head; simultaneous grouting

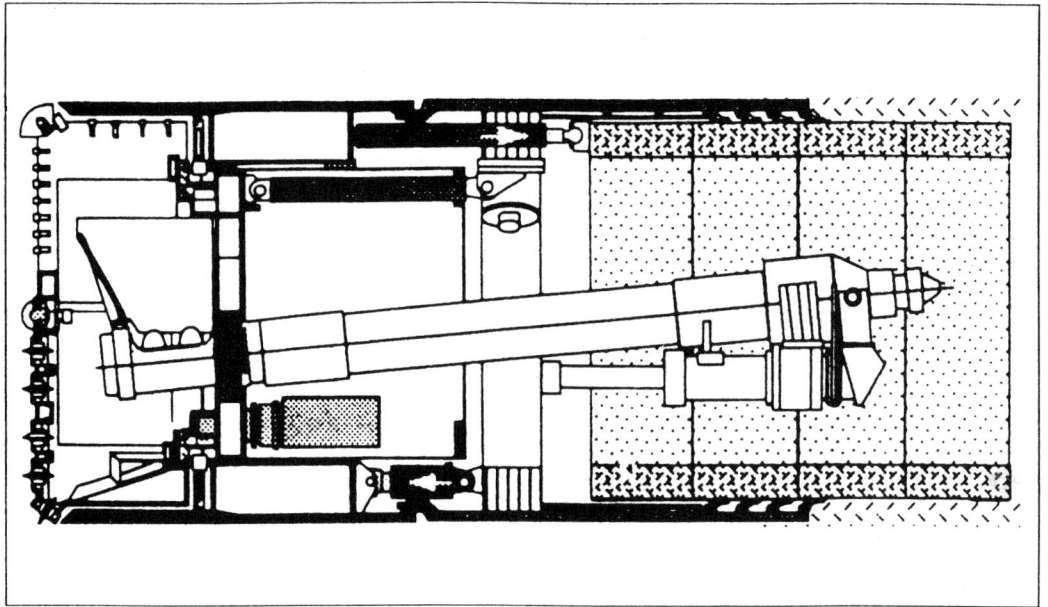

Fig. 4. *Closed mode, phase 1: retraction of front thrusters and grippers (change from open mode to closed mode); steering adjustment; application of rear thrusters*

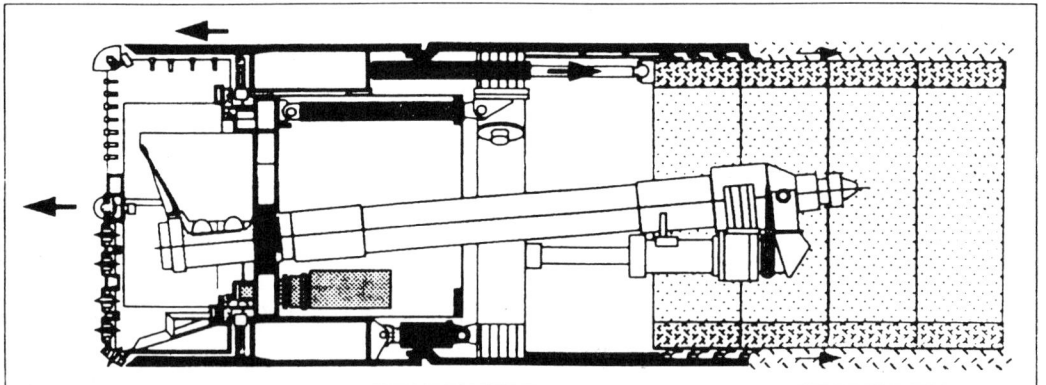

Fig. 5. *Closed mode, phase 2: start of boring; advance of telescoping head and shield under action of rear thrusters; simultaneous grouting*

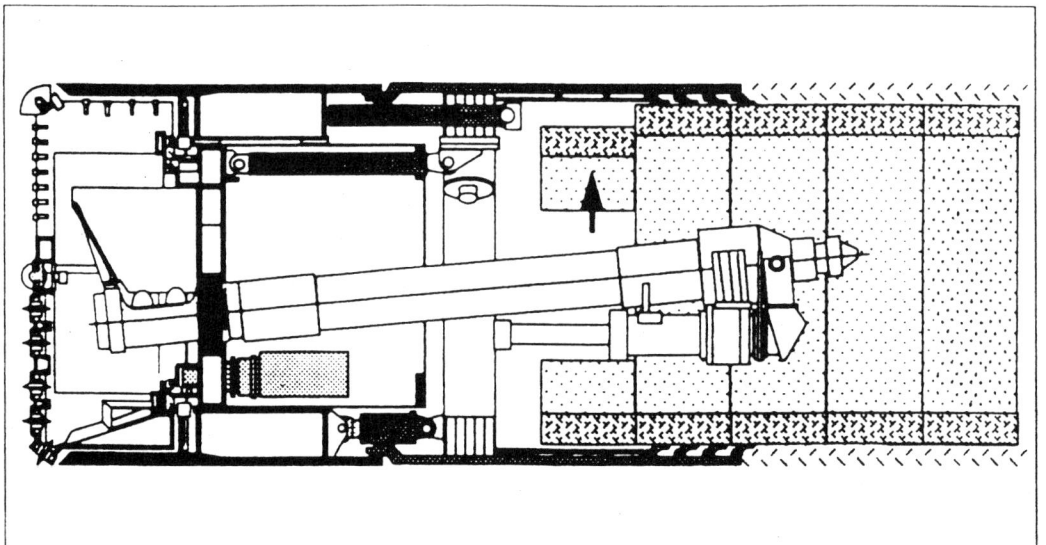

Fig. 6. *Closed mode, phase 3: retraction of rear thrusters; placing of segments*

oriented). He was thus able to judge the nature of the spoil and could alter it by adding water or changing the speed of the conveyor.

11. All movements by the TBM were recorded by the supervisory system (IBM 7532) which enabled simultaneously

(a) the main parameters of the advance to be followed in real time on a screen (forward thrust, speed of cutter head and the screws, mastic pressure in the tail seal, grouting pressure, etc.)

(b) alarms to be given in case of failure and, in certain situations, for the machine to be stopped

(c) all data about the placing of each ring to be recorded and recovered

(d) this 'ring report' to be transmitted to the surface by modem

(e) a permanent record of the progress of the TBM to be made.

12. Data were thus recorded on over 100 parameters for every 20 mm of advance.

13. The enormous amount of information collected was important for the TBM driver, and even more so for those above ground analysing the activity of the TBM and improving procedures. It also allowed preventative maintenance to be programmed.

14. Kept informed by screens, instruments, supervisor and operatives, the driver was also helped in handling the machine by a powerful computer (Alspa C350). Processing 2000 items of data, this equipment governed start-up of the machine, guidance and safety management. It also assisted in diagnosing the cause of breakdowns, more than 200 preprogrammed mes-

Fig. 7. Operating cab of tunnelling machine T3

sages could appear on its two alphanumeric screens.

15. The same equipment also operated the erector. The segments were called up, placed in the various positions, adjusted and secured, on the orders of an operator who was watching operations in the tailskin of the TBM.

16. Two other computers were installed on the support train: one (a Siemens 115U, processing 300 items of data) on the right of the front wagon to manage the injection of grout; the other (Telemecanique TSX 47-20, processing two items of data) on the twelfth wagon, to control the segment handler.

17. Some 15 numeric displays kept the operatives informed.

*Proc. Instn
Civ. Engrs
Civ. Engng,
Channel Tunnel,
Part 3:
French Section,
1994, 48–51*

Paper 10498

Tunnels—the lining

*H. Barthes, A. Bordas, D. Bouillot, M. Buzon, Ph. Dumont, J. Fermin,
J.-C. Landry, J.-P. Larive, L. Leblond, J.-J. Morlot, L. Szypura,
Ph. Vandebrouck and B. Vielliard*

■ **As the alignment of the tunnel is not
straight in the horizontal plane or in the
vertical plane, the lining had to be
designed to follow the curves. The Paper
explains how it was possible to use only
twelve positions to align the segments.
The laser and computer guidance method
is described and the importance of mix
design and placement of lining segment
mortar emphasized. The methods used to
solve problems raised by interaction
between tunnels, reinforcement of segments, nature of gaskets and bolting are
examined.**

Lining design

Between Coquilles and Castle Hill the tunnel
dips beneath the sea, climbs and drops again,
turns to the left and to the right. In short there
is no straight line. The lining therefore has to
match the curves of the alignment.

2. The tunnel lining is composed of a succession of tapered rings, 1·6 m long in the
running tunnels, a measurement which varied
by 25 mm from one side to the other (1·4 m ± 14
mm in the service tunnel). The lining consists
of two kinds of rings; 'red' ones where the
maximum length is on the right, 'blue' ones
where the maximum length is on the left.

3. For the tunnel to run straight, the red
and the blue rings had to alternate. For it to
turn left, red rings only were assembled; for it
to turn right only blue rings were placed. In the
same way, for the tunnel to descend, the rings
with maximum width in the upper half were
required. Conversely, for the tunnel to climb,
rings with maximum width in the bottom half
were needed (Fig. 1).

4. All the intermediate positions could have
permitted an infinite number of combinations
for the tunnel to undulate in plan and in elevation. In theory, only the angle of taper
limited the curvature to a minimum radius. In
practice, construction constraints—
particularly in relation to watertightness—
limited the freedom with which the rings were
positioned. The rings were bolted to each other
and the angle of rotation between two consecu-
tive rings was governed by the space between
their fixing points.

5. In the service tunnel, the axis of the key
segment could be rotated in relation to the axis
of the tunnel by +54°, +18°, −18° or −54°—
resulting in four possible positions for the red
rings and four for the blue. This provided a
minimum radius of curvature of 602 m horizontally, 572 m at humps and 926 m at hollows. In
the running tunnels, the axis of the key
segment could be rotated by +60°, +36°, +12°,
−12°, −36°, or −60°. The minimum radius of
curvature was therefore 562 m horizontally,
722 m at humps and 565 m at hollows.

6. The surveyor was required to define the
trajectory of the tunnel boring machine (TBM)
for the next 15 rings and to check all the possible combinations of rings to match this alignment as closely as possible, with optimum
centring in the tail.

7. With the succession of rings defined, a
program calculated the theoretical values to be
plotted between the ring and the tail. The
driver therefore had to guide the machine
according to the placing plan worked out by the
surveyor. If the measured values between the
lining and the TBM exceeded that planned by
more than 20 mm, a revision of the plan was
issued so that the driver could gradually bring
the machine back onto the theoretical alignment, without risk of crushing the four-brush
tail seal which ensured watertightness.

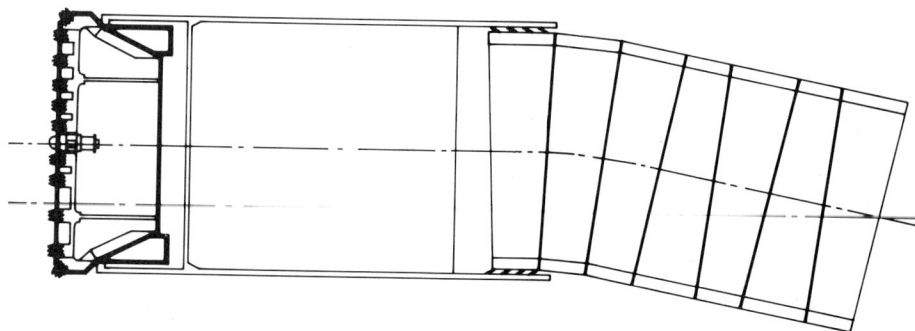

*Fig. 1. Use of tapered
lining rings to follow
the tunnel alignment*

Steering by laser and computer

8. Ring after ring, the progress of the TBM was controlled by a guidance system which, in real time, gave the driver and the surveyor the position in relation to the projected alignment.

9. In the marine tunnels, the British guidance system ZED was employed (Fig. 2). It used a target fixed to the TBM which received a laser beam from a position on the last surveying bracket behind the TBM fixed to the tunnel lining. The bearing along this light was taken by the theodolite which sent the beam.

10. Having entered the theoretical alignment in the system's computer, followed by the geometric parameters which defined the position of the laser source as well as the angle and bearing of the beam, the receptor analysed the impact of the beam on two parallel sighting targets 30 cm apart. In this way, on a given heading, the segment required to bring the TBM back on track was compared with the corresponding theoretical segment. The computer could thus calculate the variation between the TBM and the theoretical alignment. Two clinometers completed the system to indicate the roll and pitch of the TBM.

11. The divergence in millimetres of the TBM's position at the level of the target and the trend of its direction of travel were recorded continuously on the driver's monitor. The information was also printed out above ground in the surveyor's office after each ring had been placed.

12. It was thus possible to know the position of the French TBMs in relation to the UK machines at any time because they were linked geometrically to the reference points of the RTM 87 system.

13. In the land tunnels the method used to guide the TBMs between the shaft and the terminal was somewhat different. It used the German TUMA system with an automatic electronic tacheometer. This detected the position of two reflectors fixed to the TBM and, unaided, established the position of the machine by measuring the necessary distances and angles. The actual position of the TBM was compared to the position in which it should have been if it were exactly on the planned alignment. The divergence was shown on screen in the driver's cabin.

14. The TBMs proceeded on average within 3 cm of the planned alignment.

Grouting the lining

15. The rapid progress of the TBMs several kilometres away from the shaft required the development of a special type of grout to fill the annular void between the lining and the tunnel wall. It had to have the following characteristics

(a) rapid hardening to carry the first wagon of the back-up (i.e. a compressive strength of 1·5 bar after 1 h)

(b) strength after 90 days of at least 50 bar (actual average was 110 bar)

(c) stability in transport and tolerating a period of 16 h between mixing and injection.

16. The grout was made of two products, stable in isolation, with the following mix per 1000 l:

Retarded sand-based mortar
CPA cement: 82 kg
Damp fly ash (14% water content): 280 kg
Lime: 45 kg
Sand 0/4 (dry weight): 865 kg
Sand 0/1 (dry weight): 455 kg
Retarder (trisodium citrate): 2·05 kg
Superplasticizer: 3 l
Water: 275 l
TOTAL: 909 l

High alumina cement grout
Lafarge high alumina cement: 47 kg
Dry silica filler: 57 kg
Bentonite mud (30 kg/m³ water): 28 l
Retarder (citric acid): 280 g
Water: 27 l
TOTAL: 91 l

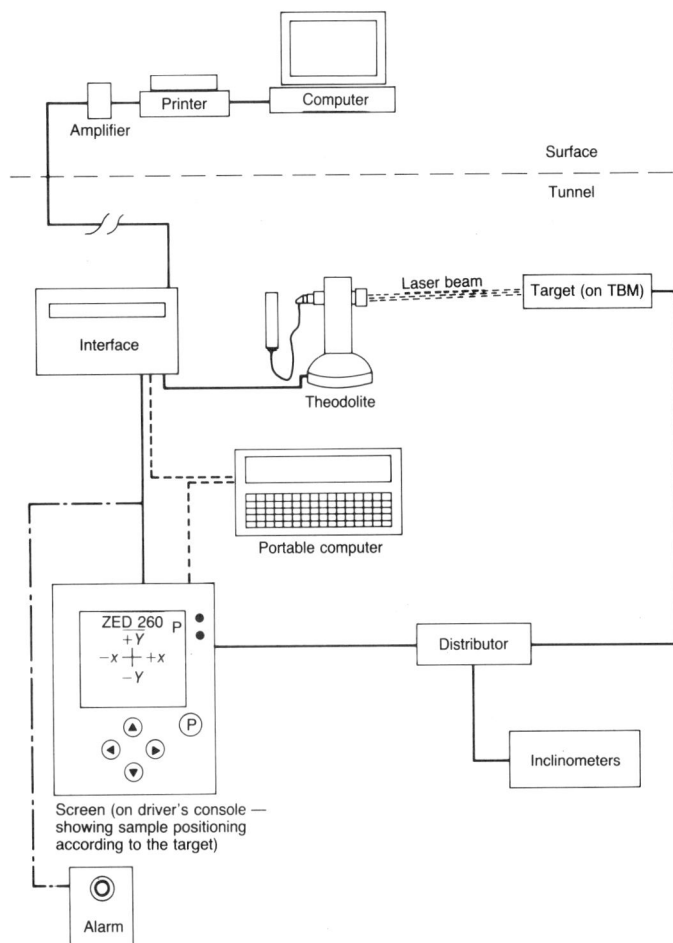

Fig. 2. ZED guidance system in the marine tunnels

These two products were prepared on site at Sangatte and stored in two silos in the shaft.

17. When each supply train was reloaded, pumps delivered enough grout for one ring from the silos to a wagon at the bottom of the shaft. In transport, the retarded sand grout was kept fluid by continuous agitation. The two products were mixed together in the leading wagon just before being injected.

18. The grout was injected between the ring and the ground starting at the bottom at the level of the invert, then in the sidewalls and last in the crown, to ensure an even filling (Fig. 3).

19. Secondary grouting was carried out at around 200 m behind the working face. A fluid bentonite cement was injected through evenly positioned holes in the segments at a pressure slightly higher than the hydrostatic pressure. This grout penetrated all the gaps between the initial grout and the ground to complete the watertight lining.

Lining design

20. Many constraints, both internal and external, had to be taken into account when designing the tunnel linings. The construction had to be watertight and capable of resisting a hydrostatic pressure which, in the deepest tunnels, could be as high as 10 bar, or 100 t/m², and have a minimum life of 120 years. These requirements led to the choice of a reinforced concrete lining, except at intersections of cross-passages with the service tunnel where cast iron segments were used.

21. The design of the lining was based on calculations according to the 'convergence–confinement' method whereby the ground and its natural stability are taken into account.

22. Two types of mathematical model were used to work out the interaction between ground and structure

(a) an analytical model by which the demands placed upon the lining of circular works were established (normal stresses, bending moment)

(b) a numerical model which addressed more difficult cases and in particular could be used to model the different geological strata and the complex constituents of the ground (plasticity, creep) for all forms of structure.

23. These methods were used to evaluate the effective pressure of the ground and the groundwater, the possible interaction between tunnels at various stages of construction, the effect of proximity to the top of the Gault Clay, and the quality of the annular grout.

24. The project required parallel tunnels to be driven with 15 m between the centres of the running tunnels and the service tunnel to accommodate cross-passages. It was essential to assess the effect of driving the two running tunnels on the lining of the service tunnel which had already been driven and lined.

25. Seven stages of execution were taken into account. The analyses provided clear proof of the importance of the nature of the contact between the ground and the lining; two hypotheses were examined—perfect adhesion and complete slippage.

26. Although the alignment is principally through the Chalk Marl the extrados of the tunnel lining is close to the top of the Gault Clay at several points.

27. A parametric study made it possible to quantify the increase in stresses depending upon the distance between the top of the Gault Clay and the axis of the structure, and upon the differences in the elasticity of the Chalk Marl and the Gault Clay (Young's modulus).

28. The effect of the proximity to the Gault Clay only became negligible at a distance from the upper axis greater than twice the diameter of the tunnel.

Fig. 3. TBM waterproofing system

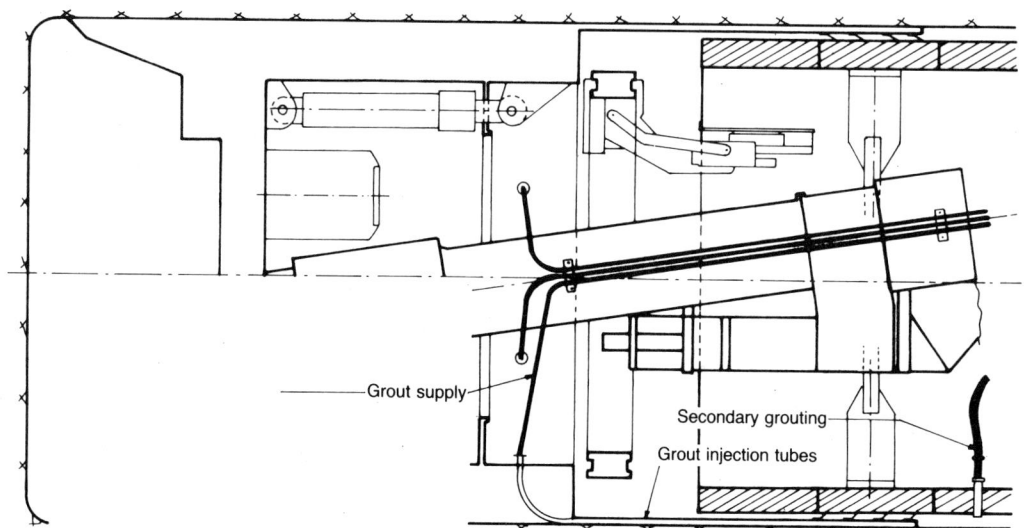

Grout supply

Secondary grouting

Grout injection tubes

29. In the case where, as on the French side, segments were placed within the shelter of the tail of the TBM, they were blocked against the ground by a ring of grout which filled the void between the extrados of the lining and the tunnel face. This non-structural element, often ignored in calculations, was of great significance. It formed, in effect, a barrier to prevent water running along the length of the lining. It was also an essential factor in the resistance of the rings to earth pressures. It provided security against pressures at the level of the abutments.

30. The rings fell into three reinforcement categories

(a) B1: standard reinforced
(b) B2: extra reinforced
(c) B3: super reinforced

(see Table 1).

31. The reinforcement consisted of cages made of Fe E40 grade high tensile transverse bars, 8–12 mm in diameter.

32. The cages were reinforced at the joints by hoops, the number of which (three to five) was a function of the strength required. The weight of the reinforcement thus varied by 15% for segments of identical dimensions.

33. Watertightness between the segments of a ring and between adjacent rings was achieved by a neoprene gasket glued in a groove which encircles each segment.

34. Two sorts of gasket were used

(a) one for segments subject to a pressure as high as 10 bar in the marine tunnels and in the land tunnels at the exit from the shaft

Table 3. Segment statistics

	Number	Concrete: m³	Steel: t	Joints: km	Tunnels: m
Service tunnel segments	79 006	94 910	7 064	675	18 435
Running tunnel segments	169 154	453 530	27 926	2 035	45 109
Total	248 160	548 440	35 010	2 710	63 544

(b) the other for segments subject to a pressure less than 4·5 bar in the land tunnels.

35. The groove for these gaskets was identical in both sorts. To obtain a perfectly watertight seal, the gaskets had to be compressed with a force approaching 8 t/m.

36. Taking into account the size of the rings (for running or service tunnels), their colour, reinforcement characteristics, gaskets and number of segments per ring, 36 types of concrete segments were manufactured for the service tunnel and as many again for the running tunnels (Fig. 4, Tables 2 and 3).

37. During the time taken to complete the grouting and block the rings against the ground, the segments were bolted together to make a ring with 22 mm bolts and to adjacent rings with 25 mm bolts.

38. The bolts, fixed in a cast-in socket, were therefore inserted as soon as the segment was placed in the tail of the TBM. They were unscrewed at the end of the back-up and reused. There were 21 holes in each ring to receive 22 mm bolts and an equivalent number of matching fixing sockets. The key segment took two bolts.

Fig. 4. Palettes of segments waiting at the bottom of the shaft

Table 1. Weight of reinforcement in the three types of segment

Weight of reinforcement according to type of ring	Service tunnel: kg	Running tunnels: kg
B1: 3 hoops	536	904
B2: 4 hoops	565	998
B3: 5 hoops	592	1041

Table 2. Main dimensions of segments

	Internal diameter: m	Thickness: m	Size of rings: m
Service tunnel	4·80	0·32	1·40
Running tunnels	7·60	0·40	1·60

Proc. Instn
Civ. Engrs
Civ. Engng,
Channel Tunnel,
Part 3:
French Section,
1994, 52–57

Paper 10499

Tunnels—the precasting plant

H. Barthes, A. Bordas, D. Bouillot, M. Buzon, Ph. Dumont, J. Fermin,
J.-C. Landry, J.-P. Larive, L. Leblond, J.-J. Morlot, L. Szypura,
Ph. Vandebrouck and B. Vielliard

■ **During the peak period, the precasting plant produced 1200 m³ of concrete per day. This Paper begins with a description of the installation, giving an idea of its extent, then looks in detail at its operations and indicates the performance levels achieved. The reinforcement shop and casting shop are described. The stringent quality control procedures, which involved certain difficulties in production, handling and storage phases are explained.**

At Sangatte the precasting plant produced 252 000 segments ahead of tunnel driving, which called for up to 3·5 km of lining in a month. Situated next to the surface installations at the shaft, the plant covered 20 ha (Figs 1 and 2).

2.   The reinforcement shop covered 12 800 m² and was split into three bays each 33 m long and 11 m high, with superstructure and crane runways weighing 620 t.

3.   The casting shop covered 10 800 m² and

consisted of four 25 m long and 10 m high bays, and 540 t of superstructure and conveyors. These two buildings were of metal frame construction with single-skin cladding, access doors for plant and runways for travelling cranes. They rested on pad foundations and were floored in concrete, with pits for casting and conduits for the utilities.

4.   The storage area under the travelling cranes was divided into five bays—two of 21 m × 180 m for the service tunnel segments and three of 27 m × 200 m for the running tunnel segments. Each bay was served by a travelling crane. Their runways consisting of 880 t of metal framework, were founded on piles. The segments were stacked on reinforced concrete beams.

5.   The concrete batching plant consisted of two mixers with an output of 60 m³/h and storage hoppers of 24 000 m³ capacity for aggregates (Fig. 3).

6.   The workshops and service buildings covered 4000 m². They consisted of metal frame buildings with single or double-skin cladding,

1. Reception of aggregates
2. Storage of aggregates
3. Casting shop
4. Reinforcement shop
5. Laboratory
6. Additive compressor area
7. Storage computer area
8. Reinforcement office
9. Concrete mixing plant
10. FT 1
11. Concrete plant foreman's hut
12. Quality control
13. Storage of glue and oil
14. Bicycle shed
15. Three-dimensional inspection area
16. Cast-iron segments
17. Station foreman's office
18. Security gate
19. Canteens
20. Cloakrooms and lavatories
21. Workshops
22. Storerooms
23. Offices
24. Liquid air tank
25. Cooling water point
26. Trolley tracks
27. Segment storage
28. Security gate

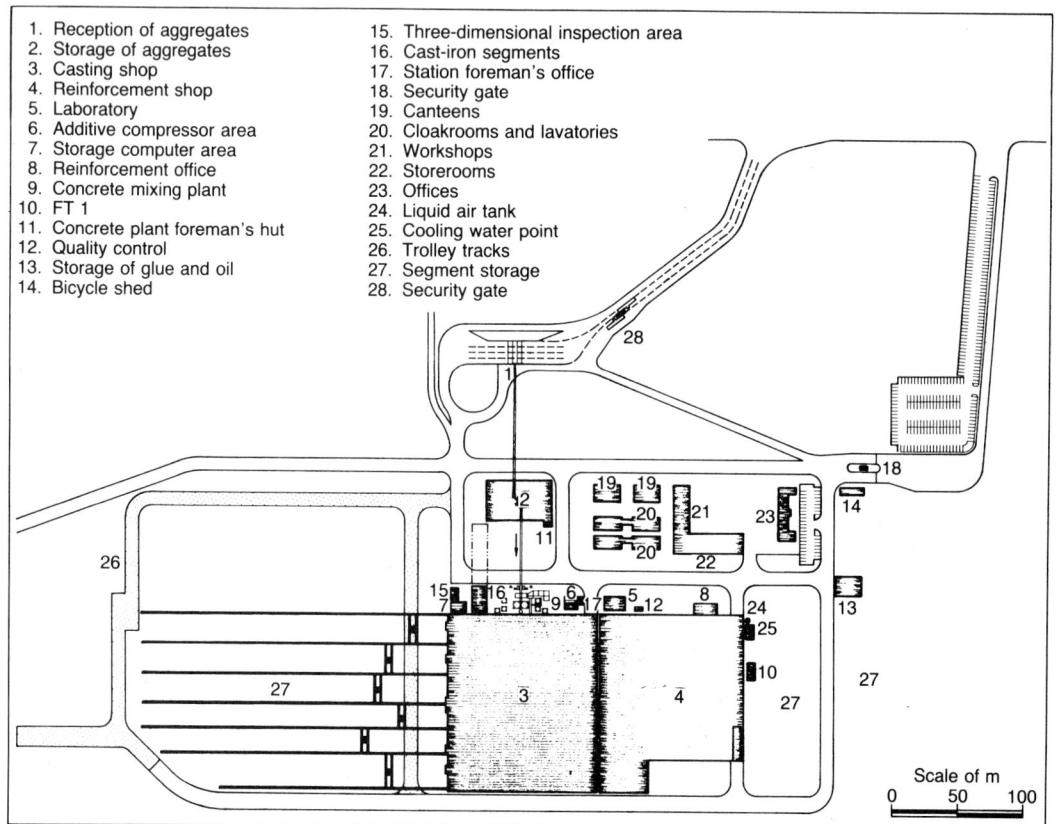

Fig. 1. Plan of precasting plant

Scale of m
0   50   100

seated on substantial foundations with concrete floors. They included

(a) a maintenance shop (800 m², 130 m² of which were offices, 5 m high)
(b) a store (640 m², 90 m² of which were offices, 6·5 m high)
(c) a glue shop (300 m², 4 m high)
(d) a compressor room and plasticizer vat (6·5 m high).

7.   The other buildings were prefabricated modular structures on concrete floor slabs. They comprised

(a) 30 offices and meeting rooms (totalling 520 m²)
(b) buildings housing the concrete testing laboratory, the segment checking area and the computer services office (total 270 m²)
(c) staff facilities and the occupational training room (1500 m²).

8.   The electricity supply to the prefabrication plant came from the 90 kV/20 kV transformer which served the whole Sangatte site. Electricity consumption rose to an average of 8000 kW/h per month during the period of full production (1989–90). It peaked at 13 000 kW/h in the cold of January 1991 and fell to a minimum of 5500 kW/h in July 1990.

Reinforcement shop

9.   To ensure quality and an even rate of production, reinforcement cages were fabricated on site.

10.   The idea of TML carrying out its own work was considered in relation to subcontracting. However, a tender from Tridamur (Davum and Mure Group) proposed a heavily automated, industrialized operation and was finally accepted. The bending schedules for the reinforcement were therefore adapted to mechanized production. The plant was handed over in working order to Tridamur who brought in their own electro-mechanical equipment. Three sheds were used, each with a travelling crane.

11.   Cutting, bending the reinforcement, and assembling the panels and the cages took place in shop 1; shop 2 housed the finishing work on the cages and initial storing; shop 3 housed the finished cages and the rehandling.

12.   The production line was organized in the following way.

(a) *Steel supply.*   Steel came from the SAM and ACOR mills. The ex-factory steel was delivered in coils and warehoused on five levels in an area of 625 m² at the entrance to shop 1 which was able to receive:
   (i) 700 t of Fe 400 steel, in 1·5 t bulk coils, of which 70% was of 8 mm and 30% was of 12 mm diameter.
   (ii) 500 t of 8 mm diameter cold-rolled wire in 1·5 t coils.

(b) *Cutting.*   The coils were unwound and the wire straightened and cut to length mechanically. The longitudinal and transverse rods were stored separately.
(c) *Production of panels.*   The longitudinal and transverse rods were taken on to the mesh production line and then welded into panels. The panels were cut to the measurements of the segments, stacked in pairs (extrados and intrados) and assembled in groups of 50. These were stored at the end of the line in an area of 1100 m² before being made up into cages.
(d) *Production of tie bars and spacers.*   An area of 250 m² was occupied by this work. It involved unwinding the coils and straightening them, cutting and making the ties and spacers.

13.   *Assembly of cages.*   The panels were delivered manually to the automated equipment which first shaped them to the required radius in a bending machine. The panels were formed in pairs. They were then hung from a monorail transport system which operated as a carousel, and were taken to the 3D/65-T welding machine. This machine received the panels, the tie bars and spacers and assembled them into cages by preprogrammed automatic welding.

Fig. 2. General view of precasting plant showing three bays for reinforcement on the right and four bays for casting on the left

Fig. 3. Concrete batching plant

53

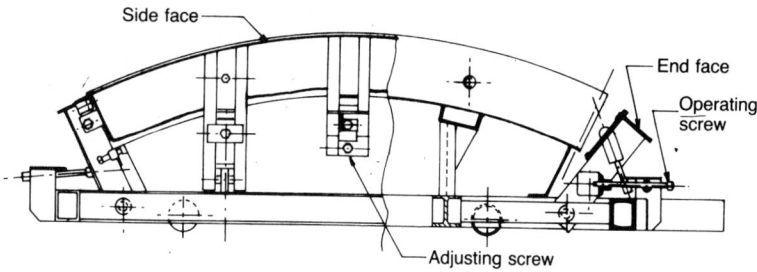

Fig. 4. Segment mould

The production capacity was 25 cages an hour, which governed the general rate of production.

14. *Finishing.* The assembled cages were taken by conveyor to shop 2 where they were distributed among five workstations, depending on their final destination. A gang of eight men operated the finishing line—cutting out for recesses, welding extra stays, and placing and welding the tie bars.

15. *Storage.* Virtually the whole area of shops 2 and 3 (6400 m²) was taken up by finished cages arranged to enable 8400 cages, or 18 days' production at peak, to be accommodated.

16. *Transfer.* The cages were loaded on to trucks in fours and moved to the casting plant. Conveyors fed the reinforcing cages to the segment reinforcement point.

Casting shop

17. Two overriding principles influenced the design of the casting plant. First, the segments were cast horizontally, extrados uppermost, to allow all the lateral faces to be cast against the formwork (Fig. 4). This had two benefits—it produced an accurate recess for the watertight gasket and high quality concrete along the longitudinal faces against which the TBMs exert pressure (Fig. 5). (Note that, in contrast, in the UK segments were poured vertically.) Secondly, the moulds were mobile, running on six chain conveyors past the workstations. Each chain conveyor had 44 mould carriers which moved on runways past the

Fig. 5. Standard segment

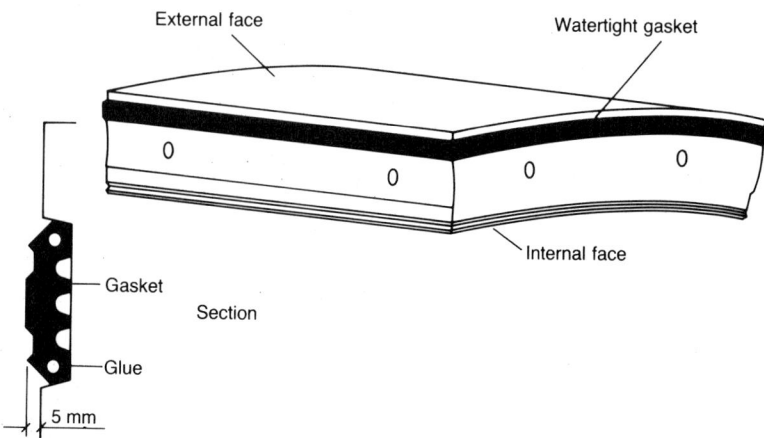

workstations. Two transfer trolleys at the ends of the track kept the chain rotating.

18. The workstations were placed along the outgoing track. The average stop time for the moulds at each station was 14·5 min. Built-in safeguards prevented the conveyor moving on before each station had completed its task.

19. Each segment went through seven manufacturing stages (Fig. 6).

Cleaning and oiling the moulds

20. Cleaning was particularly important to maintain the segment's shape and the life of the mould. The principle of 'good mould, good segment' was applied by quality control (Fig. 7). After many tests on mould oils from several suppliers without much success, an oil specially made by Pieri was selected as being most resistant to the intense vibration of the moulds, but even that was never completely satisfactory.

21. The procedure for placing reinforcement in moulds was as follows

(a) place a recess box in the bottom of the mould for the erector finger and fix it with detachable-headed screws (see Fig. 8)
(b) close the two butt joint faces and one lateral one
(c) bring the reinforcement above the mould
(d) place reinforcement spacers and put the reinforcement cage in the mould
(e) close the second lateral face
(f) position the threaded sockets for the bolts
(g) check the space left for concrete cover and carry out other general checks.

22. The reinforcement cages were supplied on flat self-propelled wagons at the upper level of the curing tunnels and stopped under the travelling cranes alongside the reinforcement positions. The supply of reinforcement cages was controlled from the workstation.

23. *Concreting the segments.* The concrete was supplied from the depot in self-propelled trucks travelling on a monorail. The mould was taken to the vibration point and the pneumatic jacks were started up. The system was isolated from the ground.

24. An initial layer of concrete half way up the mould was poured from a 4 m³ capacity hopper fitted with a vibrating chute. A mould hood for the extrados was fixed to the mould by hydraulic rams. Filling continued to the top through three screened openings in the hood (Fig. 9). The mould was vibrated in two stages; first at a frequency of 75 Hz when placing the concrete, then at 90 Hz to remove air bubbles. The hood was then unbolted and removed, the vibrating beams were then lowered to rest and the form freed.

25. *Extrados finishing.* It was necessary to finish the concrete of the extrados after the pour as the hood left an imperfect surface. The

mould was moved to the next workstation where, first, the measurements were checked using the lateral faces of the mould as a gauge. The work was particularly laborious as the concrete was very stiff after the intense vibration.

26. Next, the extrados was equalized using a pneumatic trowel, then smoothed using a Dutch trowel. The bolt socket retainers were withdrawn, waxed and replaced in readiness for their removal after the concrete was cured. The outside of the mould was cleaned. All these operations were manual.

27. *Segment curing.* The moulds passed into an insulated high humidity tunnel where they were kept for seven hours. The aim was to accelerate setting of the concrete and to make it strong enough to withstand removal of the mould and handling. The treatment took one hour at normal temperature (20°C), half an hour rising to 30°C, four hours at 30° and half an hour returning to ambient temperature. The curing tunnels had a double track for two lines of mould transporters. They were heated by electric elements and the steam was produced by small boilers.

28. *Millimetric tolerances.* When the concrete had set, the measurements of the extrados were checked, the bolt socket retainers removed, the blanks for recesses around the bolts between the segments removed, and the faces cleaned.

29. Removal of the shuttering and of the segment was achieved using a travelling crane with a spreader bar fitted with suction pads (Fig. 10). This was under manual control and the various positions were fixed previously. The crane could only traverse. The mould was positioned automatically, giving the segment the position it would keep for all future automatic movements until it was palletized.

30. The segment was then placed on an inspection bench so that the blanks could be removed and to allow an overall check. It was either passed, sent for repair or scrapped, and was marked accordingly, together with its identification number allocated by the central management computer. The segment was then placed on the exit line by the travelling crane.

31. The exit lines were installed in the bay next to the outside storage area. They consisted of six workstations. The first four were dedicated to preparing the recess for the watertight gaskets, gluing and placing them in position using a press with pneumatic rams (Fig. 11). Wooden stacking spacers were glued on to the extrados. The fifth workstation was a waiting area where the segments were sorted into those which could be stored directly, those needing repair and rejects.

Disadvantages of automatic storage

32. Because they were fitted with a large number of safety devices, the travelling cranes

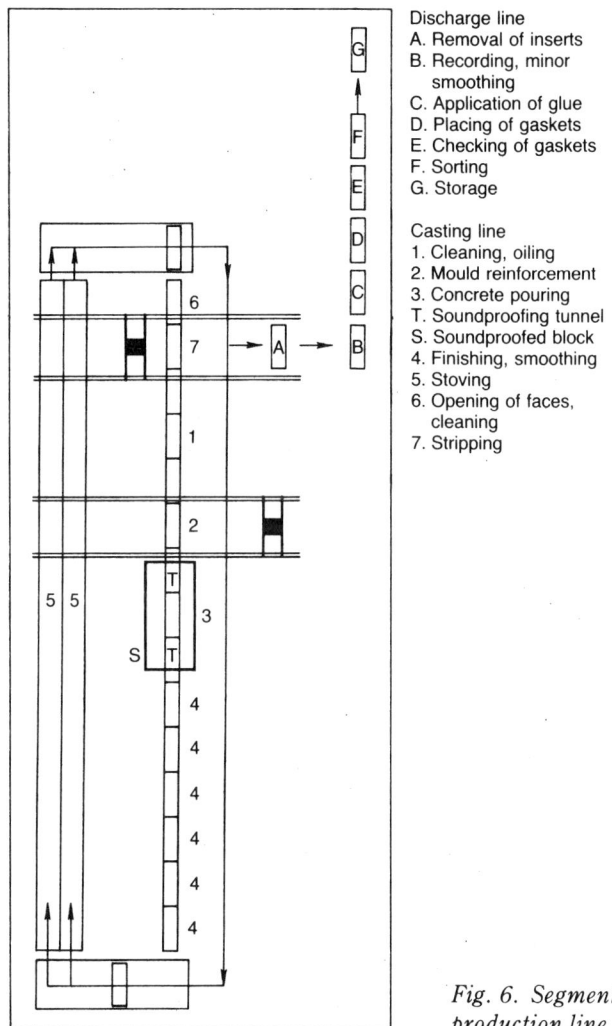

Discharge line
A. Removal of inserts
B. Recording, minor smoothing
C. Application of glue
D. Placing of gaskets
E. Checking of gaskets
F. Sorting
G. Storage

Casting line
1. Cleaning, oiling
2. Mould reinforcement
3. Concrete pouring
T. Soundproofing tunnel
S. Soundproofed block
4. Finishing, smoothing
5. Stoving
6. Opening of faces, cleaning
7. Stripping

Fig. 6. Segment production line

Fig. 7. Moulds before reinforcement

Fig. 8. Fixing reinforcement cage in mould

Fig. 9. Pouring concrete

Fig. 10. Stripping

ments had to be unloaded and the rings reconstituted separately using fork-lift trucks.

33.   Moreover, when the conveyors reduced the prefabrication cycle time it was not possible to cut the cycle time of the travelling cranes. Priority was therefore given to stocking to the detriment of loading, hence the need for extra handling operations.

34.   Finally, for handling palettes at the shaft and in the tunnel, the tolerance on size of the packed palettes was about 1 cm. The storage and automatic loading system could not cope with this degree of precision in 60% of cases. The palettes therefore had to be repacked in the open storage areas nearby (Fig. 12).

35.   With about 90 people working round the clock, the stocking and loading service provided the tunnels with deliveries of up to 120 rings a day.

High strength concrete

36.   Development of the concrete used for prefabricated segments required a long series of tests at CSTB (the French Building Research Station) under the control of Bureau d'Etudes Tunnel (BETU). A material was needed which was workable when filling the moulds, impermeable and strong (with a water/cement ratio between 0·33 and 0·35) when in position in the tunnel.

37.   Two types of concrete, B45 and B55, were used at first. By the end, to simplify the programming and avoid the inclusion of fly ash, a B45 alone was specified with the following composition for 1 m³

(a)   400 kg CPA55 PM cement from Dannes or Lumbres
(b)   140 l water
(c)   325 kg round sand (from the Oise valley)
(d)   320 kg crushed sand LS 04 from Boulonnais quarry
(e)   250 kg crushed gravel 3/8
(f)   1060 kg gravel 5/12
(g)   7 kg Sikament HR 410 or 5·5 kg Chryso Durciplast plasticizer.

38.   The concrete was mixed at a highly computer controlled site, with two twinned units (manufactured by BETP) each of 60 m³/h capacity. The aggregate was supplied, weighed and discharged into four hoppers, each capable of holding the contents of two 25 t lorries. It was then forwarded by conveyor belt to four covered vertical 24 000 m³ silos—equivalent to 10 days' production.

39.   Other conveyors took it to vertical silos above the mixers. All the weighing machines were manufactured to commercial standards of accuracy. The cement was delivered in bulk by lorry and stored in four 500 t silos representing five days' production.

40.   The mixing water was heated, depending on the weather, to a maximum of 80°C in

were susceptible to failure, which on occasion stopped the production lines! Several other difficulties with the automatic stocking system also arose. It required homogeneous production from each production line in order to deliver a complete set of six segments to make a ring, otherwise it became impossible to load the rings on to their palettes. Therefore the seg-

order to obtain concrete of about 20°C, irrespective of the ambient temperature. The 350 kW water heater had a capacity of 30 m³.

41. Consistent concrete plasticity was difficult to obtain despite maintaining the hygrometry of the aggregates to within close limits. The hygrometers mounted on all the storage containers were not reliable. Even a variation of 0·6% in the moisture content of the gravel was enough for the concrete to exceed its narrow tolerance limits. This could only be confirmed when it came to be poured. Better communication between the operator of the concrete mixing plant and the users of the concrete enabled this situation to be improved.

42. At peak periods, the precasting plant produced 1200 m³ of concrete a day. In all, 225 500 t of cement, 425 000 t of sand and 744 000 t of gravel were used to produce 563 400 m³ of concrete (Table 1).

Quality control

43. The quality of the segments was checked at all stages of manufacture by production staff according to written procedures which defined basic tasks and checks. These procedures covered all the activities from receipt of raw materials to delivery of the segments.

44. To verify that correct procedures were adopted, and particularly to monitor internal checking of production operations, independent quality control staff were present on all production lines.

45. As well as standard checking of materials in the concrete mix and carrying out crushing tests on samples, the laboratory was also responsible for

(a) checking the manufacturing parameters (vibration speeds, curing)
(b) systematically checking the measurements of the moulds every two or three weeks and each time a mould was repaired
(c) checking segment measurements where there was doubt (about 2% of production) was carried out by a machine which sensed points in three dimensions and calculated the distance between them. The moulds were measured every week (i.e. every 20 uses) by micrometric gauge, and the segment permeability checked weekly by core boring.

46. Concrete losses stabilized from between 6% and 10% when the production lines started up, to about 1% from July 1989 until the beginning of 1991. The overall losses amounted to 2%, less than the 2·5% initially forecast. However, the rejection rate of segments was 1·7%, compared with the 1·2% predicted by specialists at the outset.

Fig. 11. Gluing joints

Table 1. Performance, production and storage

Production schedule	Service tunnel	Running tunnels
Length required: m	18 434·80	45 109·10
Working days	754	837
Days of production	739	832
Days of stoppage	15	5
Days of stoppage as percentage of production	1·99	0·60
Average daily production including stoppages: m	24·45	53·89
Average daily production: m	24·95	54·22
Maximum production in one day: m	62·50	153·60
Maximum production in 7 days: m	267·40	619·50
Maximum production in 30 days: m	1 024·30	2 552·50
Maximum production in 1 week: m	258·30	598·90
Maximum production in 1 month: m	1 000·30	2 329·30

Fig. 12. Packing palettes

*Proc. Instn
Civ. Engrs
Civ. Engng,
Channel Tunnel,
Part 3:
French Section,
1994, 58–62*

Paper 10500

Tunnels—topography

*H. Barthes, A. Bordas, D. Bouillot, M. Buzon, Ph. Dumont, J. Fermin,
J.-C. Landry, J.-P. Larive, L. Leblond, J.-J. Morlot, L. Szypura,
Ph. Vandebrouck and B. Vielliard*

■ **Thanks to the fine work done by the surveyors and to the accuracy of the instruments, the tunnelling machines, which began 38 km from each other, met face to face with a deviation of 358 mm in plan and 58 mm in elevation. The Paper describes the difficulties encountered in working out a common reference system for the British and the French. It then points out the very small error margins allowed. Measurements were carried out by means of American satellites, gyrotheodolites and with the use of computer systems. Measurement accuracy was particularly remarkable if one considers the rugged conditions on a civil engineering site which is also underground.**

Alignment accuracy

Starting 38 km apart on each side of the Channel with an alignment that took them through inclines, declines and curves, the TBMs came face to face, thanks to the high precision work of the surveyors.

2. How were two TBMs, one starting from Sangatte, the other from Shakespeare Cliff, going to meet face to face having driven 38 km under the sea? This was the question often asked at the start of the works, including by cartoonists. It is true that there was no tunnel in the world as long, driven simultaneously from both ends.

3. The construction of the tunnels in the right line at the first attempt was essential because it was not possible for the tunnelling machines to reverse.

4. The accuracy achieved was remarkable. The running tunnels were driven within less than 15 cm of the design alignment, and 99% within 10 cm. This was the result of surveys of some 3500 cross-sections by laser to a precision of 5 mm. The three land TBMs finished less than 2 cm from the initial alignment. The first surveyed breakthrough under the sea between France and the UK by the service tunnel was completed with errors in alignment of 58 mm vertical and 358 mm horizontal, compensated for in the last 100 m of the drive.

5. The degree of accuracy demanded of the surveyors was extremely high. It was up to them to position the TBMs correctly at the start, on each side of the Channel, and to show the path to follow in order to meet under the sea or to arrive at the centre of the Portals of Beussingue and Castle Hill, built in advance.

6. Points of reference were first chosen in France and England from among the existing geodetic surveys, marked on the ground by triangulation pillars and benchmarks. Their coordinates had been calculated by the Transmanche common system of reference with the help of geodesic surveys and astronomical sightings.

7. Next, secondary reference points were fixed, closer together, starting from the geodetic points on the surface, by triangulation and refined topometric techniques.

8. At the bottom of the tunnels, points of reference were established as the TBMs advanced.

9. Therefore it was the surveyor, navigating without visual reference, who gave the TBM driver all the indications required to drive and instal the lining as close as possible to the planned alignment.

10. The major difficulty lay in the distance of 38 km which separated the two points of access to the tunnel workings. Between them lay an obstacle even more difficult than a mountain—the sea, which prohibited any fixed station and excluded the usual surveying methods practised on land.

11. Other constraints also influenced the choice of methods and topographic instruments. The construction programme was very short. The alignment was irregular because it had to satisfy geometric criteria (connected with high speed trains), geotechnics and geography.

12. Also, the diversion of the service tunnel in the zone of the crossover and before the exit at the Beussingue Portal, with radii of curvature reduced to 2000 m, shortened the surveyors' lines of sight and could therefore affect the accuracy of their surveys.

13. In all, however, the civil engineering tolerances of about 15 cm which had to absorb the errors inherent in the guidance and the handling of the TBMs, the scheduling of the succession of rings, the manufacture, placing and deformation of the segments, left the surveyors with a margin of error of only 20 mm within which to work.

14. To obtain a result of such precision without slowing the progress of the tunnel drives, TML employed its own in-house survey team.

A common map

15. The team's first task was to harmonize the approach on each side of the Channel. To

implement the project between the UK and the Continent it was essential to agree on a common system of projection, which would make it possible to establish a common map. However, a compatible topometric reference system does not exist between the two sides of the Channel.

16. With regard to planimetry, the Mercator projection reigns in the UK, the Lambert in France. Even their altimetric systems cannot be precisely correlated. Tidal gauge measurements from 1958 imply a discrepancy of 44 cm between the zeros of the French and British levelling systems.

17. A system of common coordinates in three dimensions (x, y, z) was therefore established at the start. It used the Mercator projection but originated from meridian $1°30'$ E lying at the midpoint of the tunnel to minimize distortion (Fig. 1).

18. Five geodetic points were chosen for their stability and their position between Gravelines (top of the water tower) and Boulogne (some benchmarks fixed to the bunkers of the last war) and seven points between Hastings and Ramsgate (Fig. 2).

19. In 1987 the Global Positioning System (GPS) was selected. This method, which made use of the potential of 18 Navstar satellites, made it possible to overcome the atmospheric conditions of the North Sea which affect optical sightings and to obtain the relative precision of 10^{-6}, or about 5 cm over a distance of 50 km, between reference points.

20. A few nights in October 1987 were sufficient for the signals emitted by the atomic clocks in the US satellites to be received by the French (Sercel), British (Trimble), and US (Texas Instruments) receivers stationed at the twelve points.

21. The coordinates of the twelve base points thus gathered formed the in-house basis of Transmanche 1987 (RTM 87) by which all the planimetric measurements were subsequently expressed.

22. There remained the orientation of the figure formed by the twelve reference points in relation to the North. Several more nights were necessary to sight the pole star from three geodetic points at places in the Pas de Calais. With the help of powerful theodolites and precise clocks, the azimuth of the basis for calibrating the gyrotheodolite was determined. This instrument, a form of gyroscope (a 'top' spinning at 4200 r/min aligning with the axes of the poles) indicated the geographic North and was of great importance in aligning the tunnels. Accurate timekeeping was essential.

23. In order to have points of reference of direct use in setting out the access shaft, the Beussingue Portal and the site installations, the available geodetic and altimetric information had to be expanded, so benchmarks more than 5 km apart were selected.

24. The triangulation allowed additional points of reference to be positioned on site. Thanks to concrete pillars, fitted with a centring system for setting in place the theodolite and the infra-red rangefinder, it was possible to determine the coordinates of reference points separated by one or two kilometres by measuring the angles and the distances. Some 100 surface points were thus defined on both sides of the Channel by the RTM 1987 system.

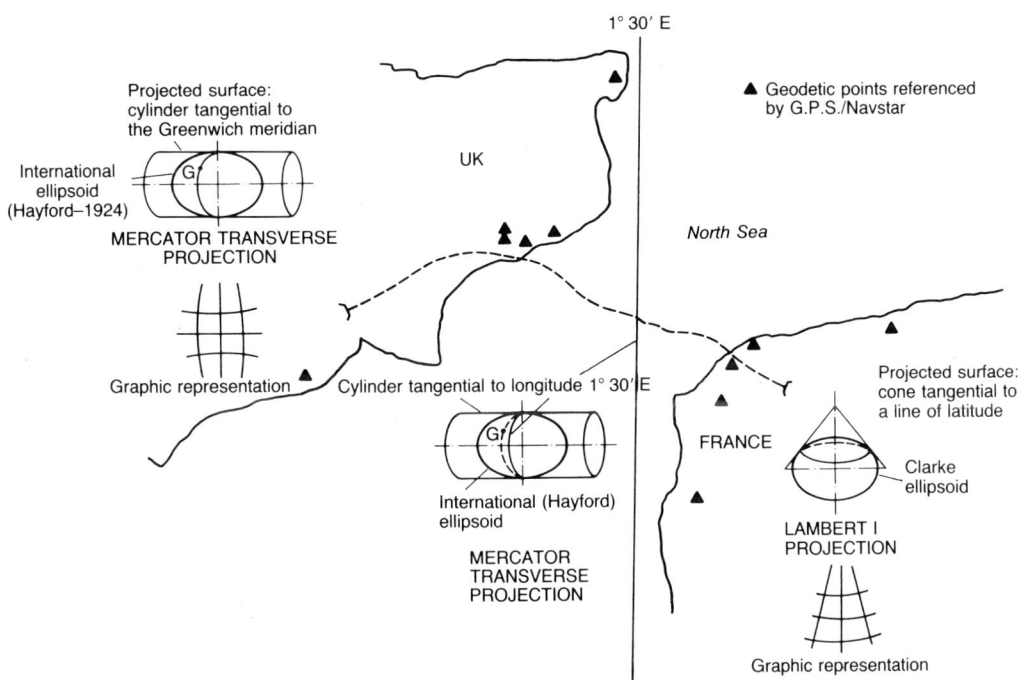

Fig. 1. On the British side, the Mercator projection; on the French side the Lambert projection; and, to reconcile the two, a TML system

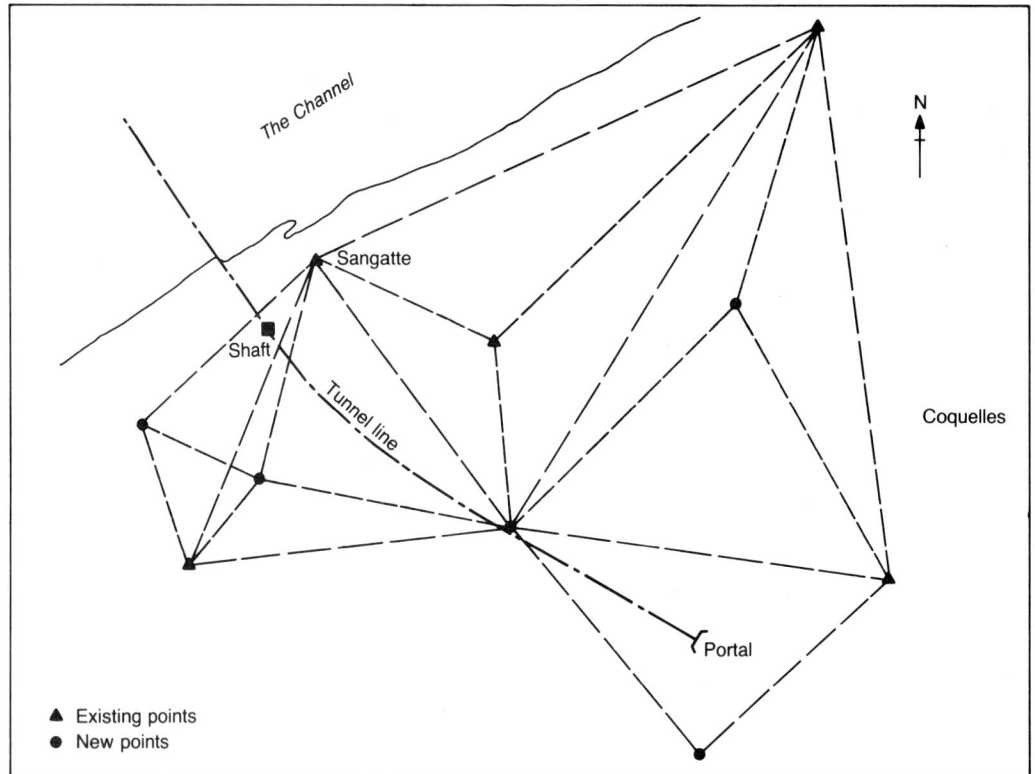

Fig. 2. The Sangatte points

Geometry inside the shaft

25. Step by step new reference points were carefully fixed as the shaft was constructed. The geometry was transferred to 45 m below ground, at the level of the marshalling chambers, by means of eight tubes on overhangs at the top of the shaft and eight corresponding brackets fixed to the ceiling of the marshalling chambers vertically below the tubes (Fig. 3). This lowering of the base was achieved with a precision of 0·6 mm.

26. In fact, the maximum base length in the shaft was 54 m. This had to suffice for the orientation of the tunnels. A simple error of 5 mm at either end would have had direct repercussions on the orientation of the TBMs and would have represented differences of 30 cm at the portal, and 1·45 m at a point 15 km under the sea, even supposing that the geodetics and subsequent underground topometry could be perfect.

27. No error could therefore be tolerated. For that reason the surveyors added to the measurements of angles and distances, that of azimuths—angles measured in relation to geographic North using a gyrotheodolite, which allowed consistent orientation in the tunnels.

From module to module

28. As the TBMs advanced, metal brackets and reference points were fixed every 50 m to the tunnel lining working from the back-up.

29. Observations between these points using topographic instruments mounted on the brackets, after calculation and compensation for error on computer, gave the RTM 1987 coordinates of the ten points which made up the topographic module (Fig. 4). Thus, gradually the modules formed a long chain from the shaft to the face.

30. From these figures of a base of 400 m, the length of all the sides was measured to within a millimetre, all the angles measured with exceptionally fine precision, and two azimuths measured by gyrotheodolite, the latter having the advantage of eliminating the deceptive effects of atmospheric refraction (Fig. 5).

31. Before and after each azimuth measurement in the tunnels, the gyrotheodolite had to be recalibrated to the base established on the surface. This instrument was very sensitive to vibration, wind, knocks, temperature differences. It was often necessary to let it stand for at least half an hour before taking readings to allow it to adjust to the ambient temperature. In the open air, to protect it from the wind, a shelter had to be built on top of the bunker chosen as a calibration point. Even underground it had to be handled with kid gloves (Fig. 6).

Two bases for recalibration

32. On the French side, the distance measuring instruments were regularly recalibrated on

a 300 m base line which was constructed under temperate climatic conditions. Using seven aligned pillars, these measurements were precise to 0·06 mm using Invar wire. As the project was linear, the surveyors were not particularly concerned about small errors resulting from the system which could build up during measuring each 200 m side of the polygon.

33. They were more worried about the soundness of the orientation of the UK base which was used almost every day to calibrate the gyrotheodolite. It was only 200 m long and it only needed the coordinates to be out by 5 mm at either of the two ends for this small error over 200 m to have an effect on the whole of the tunnel alignment 110 times longer. The effect of this fault in alignment at the end of the UK drive would have been 50 cm.

34. For this reason the base azimuth was verified several times against the pole star from both sides of the Channel. It was also the reason for taking the French gyrotheodolite to Shakespeare Cliff, and its UK counterpart to Sangatte several times, to check that the instruments were both registering the same values.

Method of correcting alignment

35. At all events and despite precise daily guidance, the two TBMs could not, except by extraordinary luck, end up precisely face to face. Given the methods employed, error in alignment at the meeting point should not have exceeded 50 cm laterally or 25 cm vertically. On this estimate, it was decided to stop the machines in the service tunnel 100 m apart, a distance at which it should be possible to rectify the alignment, taking account of incline and radius of curve; criteria which would not slow down future rail traffic.

36. A horizontal borehole, 100 m long and 56 mm in diameter, was therefore driven from the cutter head of the UK machine towards the French TBM. On 30 October 1990 a historic—and friendly—current of air flowed between the two countries.

37. It only remained to make a survey of the position of the French machine in relation to its UK counterpart before the latter continued the drive for a further 40 m and then changed direction towards the south and was abandoned.

38. For this operation the surveyors used a Swedish instrument—Maxibor. In tubes 3 m long and 5 cm in diameter, screwed together over 100 m, CCD cameras registered two illuminated rings positioned 3 m and 6 m in front of a measuring point. Successive deviations of the tube after each 3 m heading were registered so as to measure the levels inside the two sections of tube.

39. Interpretation of these observations by computer provided a longitudinal and plan

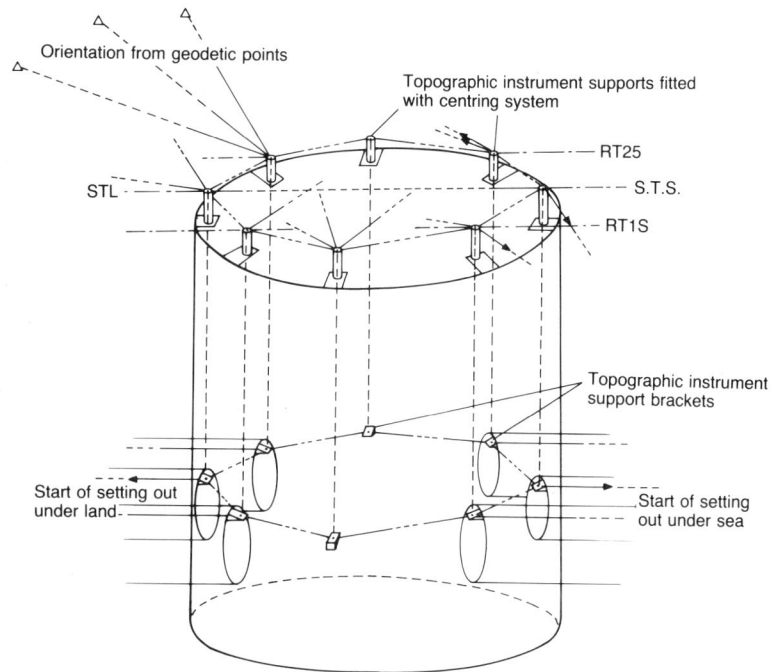

Fig. 3. Setting out in the Sangatte shaft

Fig. 4. Work in a module

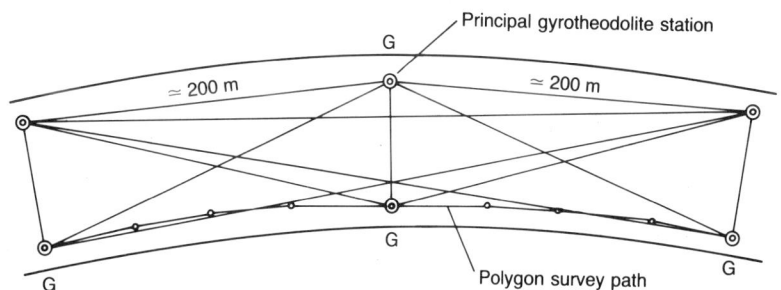

Fig. 5. Surveying module showing measurement of azimuth of a polygon side with the Gyromat gyrotheodolite

Fig. 6 (right). Sighting with the gyrotheodolite

view of this reconnaissance heading. The survey undertaken to link the end of the Maxibor with the centre of the cutter head of the French TBM completed the data, establishing that the two machines were 50 cm off line to the nearest 10 or 20 cm.

40. A makeshift water level completed the data. By means of a plastic pipe filled with deoxygenated water slid into the borehole tube, the atmospheric pressure at the two ends was read. Direct comparison of the meniscus to the UK and French reference levels indicated a difference in level of 8 cm, to within 2 or 3 cm.

Fig. 7 (below). Marine service tunnel showing the junction between the French and English surveying paths

Topographic junction

41. A further month had to pass after the French service tunnel TBM had been completely dismantled. On 1 December 1990 a small heading was driven between the two sections of tunnel to unite the two topographic paths.

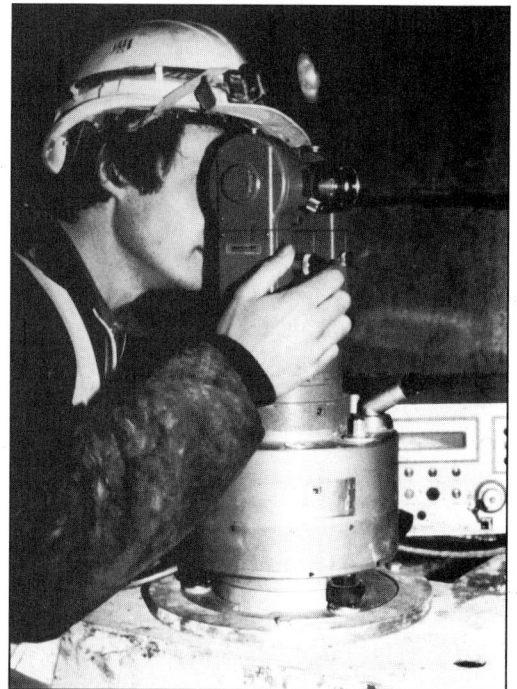

On Monday 3 December 1990 the two British and French survey teams also compared their measurements and found that over 38 km of drive, the precise gap between the two TBMs, was 358 mm in plan and 58 mm in elevation (Fig. 7).

42. These figures conformed to the estimates of a month earlier. The UK TBM was easily able to counter this discrepancy by following a corrective line which conformed to the specifications.

43. The breakthrough of the service tunnel provided the surveyors with a unique reference line between Coquelles and Folkestone. Great Britain was no longer an island.

44. It then remained to relate the geometry of the running tunnels to that of the service tunnel, using transverse links provided by the cross-passages. These loop connections eliminated the need for horizontal reconnaissance probes between the working faces. The French TBMs could be guided directly to the final rings placed by the British team.

45. The errors in alignment between the cutter head of the French machines and the final British rings were only 25 mm in the north running tunnel, and 35 mm in the south running tunnel. Not bad for meeting points set 40 km apart separated by the sea!

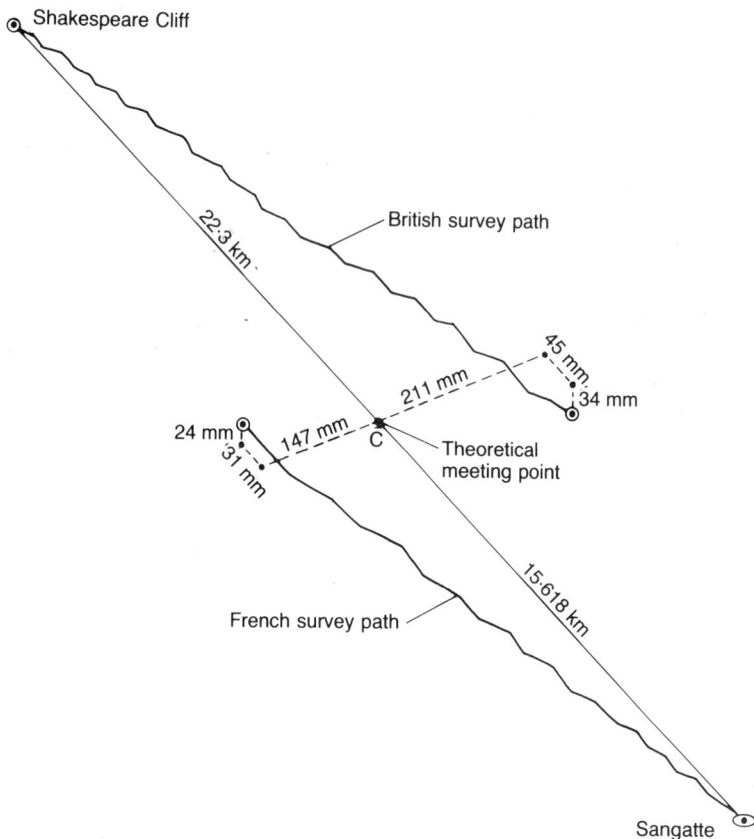

Shakespeare Cliff

British survey path

22.3 km

211 mm

45 mm

34 mm

24 mm

147 mm

31 mm

C

Theoretical meeting point

15.618 km

French survey path

Sangatte

Tunnels—special works

H. Barthes, A. Bordas, D. Bouillot, M. Buzon, Ph. Dumont, J. Fermin,
J.-C. Landry, J.-P. Larive, L. Leblond, J.-J. Morlot, L. Szypura,
Ph. Vandebrouck and B. Vielliard

Proc. Instn
Civ. Engrs
Civ. Engng,
Channel Tunnel,
Part 3:
French Section,
1994, 63–75

Paper 10501

■ In this Paper, the special works to tunnels
are considered in three parts: the transverse passages, the pumping station and
the railway crossover. The transverse
passages are of two types: cross-passages
proper and piston relief ducts which allow
the exchange of air masses. The Paper
describes how both were constructed. The
pumping station, designed to dewater the
tunnels, was of conventional design; the
Paper deals essentially with the civil
engineering of this structure. The
railway crossover allows the linking of
the north and south lines in a cylindrical structure of constant section. The
different phases of its construction are
described.

Transverse passages

On the French side 187 cross-passages and
piston relief ducts connecting two or three
tunnels were driven. Their profiles varied
according to end use.

2. The French and the British each constructed the special works next to their own
part of the tunnels. In the central section where
the French running tunnels run each side of the
service tunnel driven by the British, the cross-
passages and piston relief ducts were executed
by both teams.

Cross-passages

3. To provide connections between the three
tunnels, cross-passages were excavated at
about every 375 m (Fig. 1). This interval was
chosen with regard to the 750 m long shuttles
so that there are always two cross-passages
accessible from a stationary shuttle. In particular they are intended to

(a) allow passenger evacuation in the event of
an incident in a tunnel
(b) give access to the running tunnels from the
service tunnel for maintenance
(c) provide a route for supply services from
the service tunnel
(d) provide tunnel ventilation.

4. The cross-passages are usually in pairs
facing each other on opposite sides of the
service tunnel. They were designed to provide a
continuous level floor, without steps, between
the platform of the service tunnel and the emergency walkway of the running tunnel, irrespective of the level of the latter:

5. Standard cross-passages have the following dimensions

diameter: 3·3 or 4·8 m
length: 9·93 m per branch passage
difference in level between each end: 0·35 m
downhill gradient: 3·53% on average.

6. The lining was either cast iron segments
(0·7 m wide, maximum weight 100 kg) or unreinforced concrete 22 cm thick placed in situ.
7. The passages are all fitted with fire doors
close to the junction with the running tunnel.
8. On the French side, 57·5 cross-passages
of 3·3 m diameter were constructed (including 9
special purpose and 10·5 multi-purpose), and 2
of 4·8 m diameter.

Piston relief ducts

9. Between the two running tunnels, the
piston relief ducts ensure a constant aerodynamic state by allowing the exchange of air
masses and equalizing pressure increases and
decreases caused by passing trains (Figs 2
and 3).
10. The design of these ducts met the following requirements

(a) no angles which would generate significant
aerodynamic losses
(b) openings to be compatible with the results
of aerodynamic tests

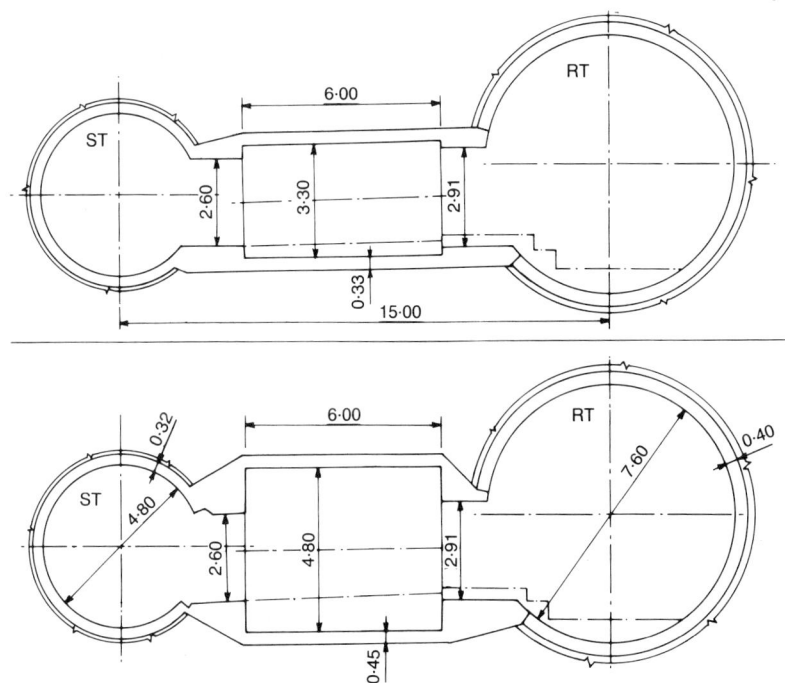

Fig. 1. Sections of
cross-passages
(dimensions in m)

Fig. 2. A completed piston relief duct

(c) air current should arrive as low as possible to minimize the stress on the shuttle

(d) the cross-section of piston bore necessary is about 12 m² per kilometre which corresponds to a 2 m diameter duct every 250 m. Aerodynamic testing at the Saint-Cyr laboratory showed that this spacing was necessary with ducts of 3·14 m² cross-section.

11. The standard duct has the following characteristics

length: 23·33 m
radius: 18 m
height between crown of duct and level at entrance: 3·41 m
downhill gradient of the duct to the entrance to the running tunnel: 27%
thickness of the chalk between the extrados of the duct and of the service tunnel: 1 m.

Fig. 3. Half longitudinal section of a piston relief duct (dimensions in m)

12. The duct opening in the tunnel is fitted with a round piston relief damper of 2 m dia., normally open, situated alternately on the north and south sides so that the tunnels can be isolated from each other.

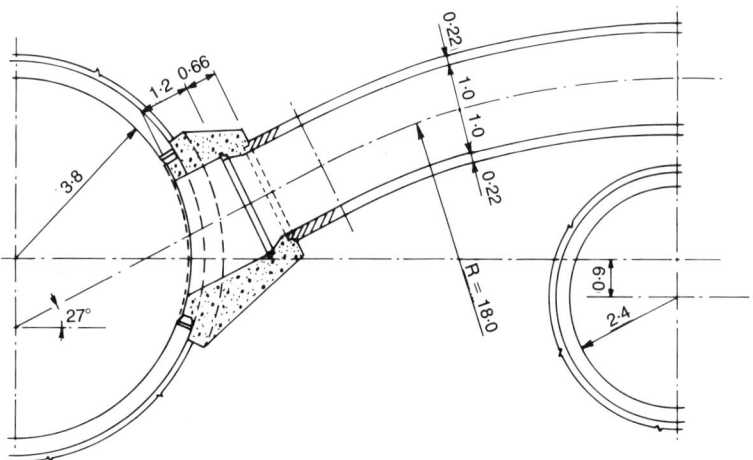

13. On the French side, 85·5 piston relief ducts were built; of these 10 were specials and 15·5 were in collaboration with the British.

Electrical equipment rooms
14. Different types of passage were constructed to house equipment supplying the system with electricity (power, lighting etc.).
15. On the French side these works include

(a) four electricity substations of 4·8 m dia.
(b) two housings for catenary equipment of 4·8 m dia.
(c) 27 passages for electrical equipment: five of 4·8 m dia. (one special purpose) and 22 of 3·3 m dia. (three special purpose)
(d) nine signal rooms (one special purpose) of 3·3 m dia.

16. These works were attached to the service tunnel in such a way as to provide a continuous floor between the cross-passage and the platform, with a minimum gradient for water run-off.
17. On the running tunnel side, the cross-passage opening was set in such a way that the gradient towards the service tunnel was as slight as possible, without falling below the feasibility limit for mechanical excavation.
18. The typical electrical equipment room is

length: 9·92 m
difference in level between each end: 0·36 m
average gradient: 2%.

19. The lining consists either of small format cast iron segments (70 cm long, 100 kg maximum), or of unreinforced concrete placed in situ. The concrete walls of the equipment rooms of 3·3 m dia. are 30 cm thick, and those of 4·8 m dia. 45 cm thick.

Construction
20. The same procedure was employed for the principal construction stages of the passages and ducts irrespective of the type or the diameter.
21. Given the construction sequence, the special rings for the service tunnel which cater for the intersection were 'receivers'. Their design was a function of the bearing loads at different construction stages. These rings were entirely of cast iron. They included two parts which corresponded to the finished opening and which could be dismantled in smaller elements from the extrados (so that traffic and temporary equipment involved in the construction phases of the service tunnel was not disrupted).
22. It was decided to use manual methods of excavation together with hydraulic rock breakers (Fig. 4). These methods had the advantage of simplicity; they allowed adjustment of the number of work stations, which proceeded simultaneously in order to suit production targets, and could be adapted more

easily to the different types of ducts than mechanical methods using a road header.

23. With a view to adaptability, TML chose cast iron linings. These were used for ducts of 2 or 3·3 m dia. in two sections in the south marine running tunnel and ten sections in the north marine running tunnel.

24. Owing to difficulties encountered as much in guaranteeing the supply as in resistance to corrosion, cast iron was abandoned in favour of concrete placed in situ (Table 1).

25. Two main options were investigated for openings off the running tunnels.

(a) *Dismantling*. In this instance specialist rings were mixed. In the area next to the opening the segments were of cast iron, the remainder of concrete. All cast iron parts, whether or not they were to be dismantled, weighed some 35 t per intersection. Placing mixed rings during the drive would have hampered the progress of the TBM to some degree.

(b) *Cutting out the lining*. This solution needed only about 5·6 t of cast iron per intersection. It required no special provisions to be taken into account during the drive, and the openings could be built to order in each case. This second technique was adopted. It involved boring continuous holes by diamond core drilling.

26. The five main stages of constructing a transverse passage are shown in Fig. 5 which illustrates a cross-passage with a cast iron lining.

Phase 1. Cutting an opening in the running tunnel lining by drilling through the concrete segment.
Phase 2. Bringing in excavating equipment and starting excavation (Fig. 6).
Phase 3. Excavating and placing lining. The track next to the entry is taken out of use to allow work to proceed.
Phase 4. Placing the cast iron structure forming the junction with the running tunnel; grouting the cast iron linings.
Phase 5. Continuing the passage to join the service tunnel and opening access to the parallel running tunnel if required.

27. In each of the running tunnels a series of workstations was set up. These operated in parallel and together covered a maximum length of two sections (about 2 km).

28. The teams undertook the following operations.

(a) Placing points and switches and siting a monorail next to each cross-passage.
(b) Cutting out the segments using a lifting cradle with generating set on board, two Deudian hydraulic diamond core drills of 200 mm dia.
(c) Placing anchoring rods.
(d) Dismounting concrete segments and ele-

ments of the cast iron lining using a hydraulic crane, an innovation from Semafor.

(e) Excavating passages (two to four teams). For each passage two automatic conveyor belts, two or three mini back-actors (Job or Kubota of 2·8 to 3·5 t, fitted with buckets of 68 l capacity and Montaberti 75 or 91 hydraulic rock breakers) and two EKFO shielded conveyors were used.

(f) Placing concrete (two teams) in all the passages with five CITA agitators of 5–6 m³ capacity for each passage, two Putzmeister hydraulic concrete pumps of 10–15 m³ per hour.

Fig. 4. Hand excavation of a piston relief duct

Table 1. *Quantities for transverse passages*

Transverse passage	Cast iron solution	Concrete solution
Piston relief duct 2 m dia.		
Volume of excavation	113 m³	123 m³
Cast iron lining	16 t	—
Grout	27 m³	—
Unreinforced concrete	—	31 m³
Cast iron damper	11·2 t	11·2 t
Reinforced concrete	21 m³	21 m³
Reinforcement	1 t	1 t
Cross-passage 3·3 m dia.		
Volume of excavation	96 m³	106 m³
Cast iron lining	10 t	—
Grout	14 m³	—
Unreinforced concrete	—	23 m³
Concrete for invert	23 m³	23 m³
Cast iron junction	5·6 t	5·6 t
Reinforcement	1·45 t	1·45 t
Cross-passage 4·8 m dia.		
Volume of excavation	—	197 m³
Unreinforced concrete	—	58 m³
Reinforced concrete	—	31 m³
Concrete for invert	—	10 m³
Reinforcement	—	3·3 t

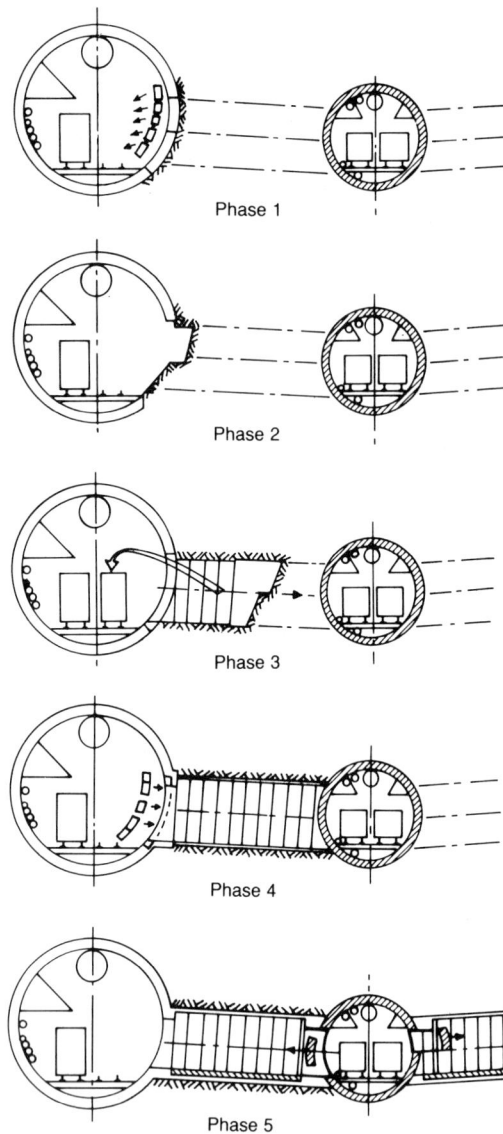

Fig. 5. Communication passage construction phases

Phase 1

Phase 2

Phase 3

Phase 4

Phase 5

Fig. 6. Driving a cross-passage with mini-shovel after cutting away the lining of the running tunnel

In addition, a lifting cradle was available for use in all the tunnels with hydraulic equipment, as well as five 650 kVA transformers and twelve trackwork switchpoint sets.

The pumping station

29. Only one pumping station was planned on the French side, situated at a low point 8·8 km from the Sangatte shaft. The two other pumping stations were in the British section.

30. The French pumping station comprises two symmetrical structures parallel to the direction of the tunnels and situated between them (Fig. 7). In this zone the distance between the running tunnels was increased to 40 m to accommodate the two shafts, the electrical equipment chambers and the emergency sumps.

31. The pumping station consists of two structures symmetrical around the service tunnel (Figs. 8 and 9) each comprising the following.

(a) A concrete shaft of 36 m² elliptical section and 11·2 m deep. The upper part of this shaft houses the pump motors, and the lower part the pumps themselves.

(b) A 4·8 m dia. passage connecting the tunnels and containing the surge tanks to relieve the water hammer.

(c) A 4·8 m dia. 45 m long chamber parallel to the tunnel and containing electrical equipment serving the pumping station and the tunnels. This chamber is connected to the tunnels by 3·3 m dia. passages.

(d) A service sump of 4·8 m dia. and with a capacity of 90 m³ at the bottom of the shaft and perpendicular to the tunnels. Both service sumps are linked to the three tunnels by vertical pipes of 400 mm internal dia.

(e) A dry room of 4·8 m dia. connects the lower parts of the two shafts, and houses the pumping equipment.

(f) An emergency sump of 3·3 m dia. parallel to the tunnels and joined to the lower part of the shaft. It is about 156 m long, with 120 m having a gradient of 0·5%. The upper part, 36 m long, has a gradient of 20% and for construction purposes is linked to the cross-passage.

32. The capacity of each of the two emergency sumps is 840 m³ of which 270 m³ are used to store hazardous liquids. This should hold the total seepage from the three tunnels over 14·5 km for two hours.

33. The average seepage expected is 6 l/s per km of tunnel. Thus the total capacity is $2 \times (840 + 90) = 1860$ m³ which corresponds to about 6 l/s per km $\times$ 7200 s $\times$ 43·5 km.

Construction

34. Ground investigations from the service tunnel and further experience gained during

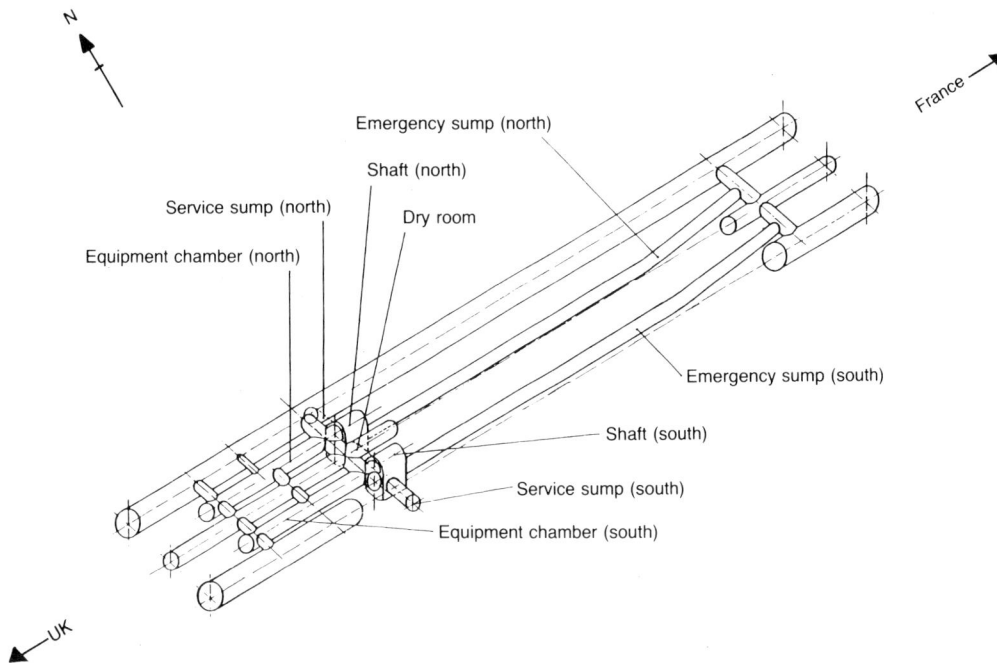

*Fig. 7. General
layout of pumping
station*

tunnel driving permitted optimum safety conditions during all construction stages.

35. Risks were reduced and flexibility achieved by using the excavating and spoil facilities from the service tunnel, whereas concrete was supplied through the running tunnels by normal means.

36. According to the simplified programme, the order of construction was as follows (Fig. 10)

(a) equipment gallery on the Sangatte shaft side and dome with temporary support of 150 mm thick shotcrete and rock bolts

(b) equipment gallery on the UK side, of 4·8 m diameter with temporary support of 150 mm shotcrete alone

(c) shaft 1st phase to level −3·88 m (level 0 is the invert of the service tunnel) with temporary support of 150 mm shotcrete and rock bolts

(d) emergency sump 3·3 m dia. with temporary support of rock bolts

(e) shaft 2nd phase to level −8·18 m with temporary support of 150 mm shotcrete and rock bolts

(f) service sump of 4·8 m dia. with temporary

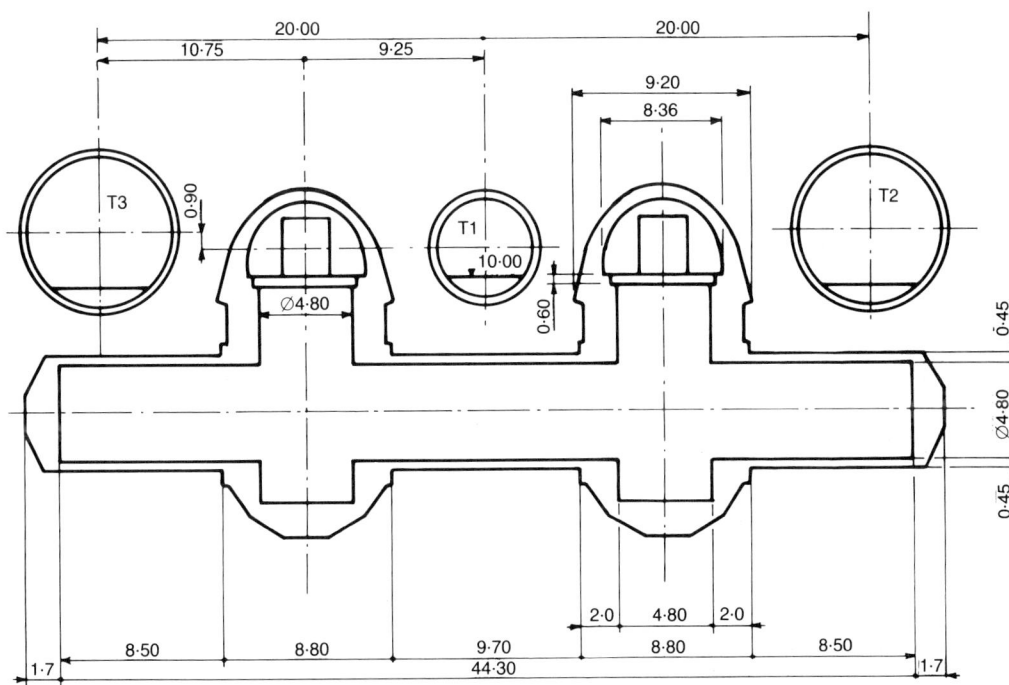

*Fig. 8. Section of
pumping station on
service sump
(dimensions in m)*

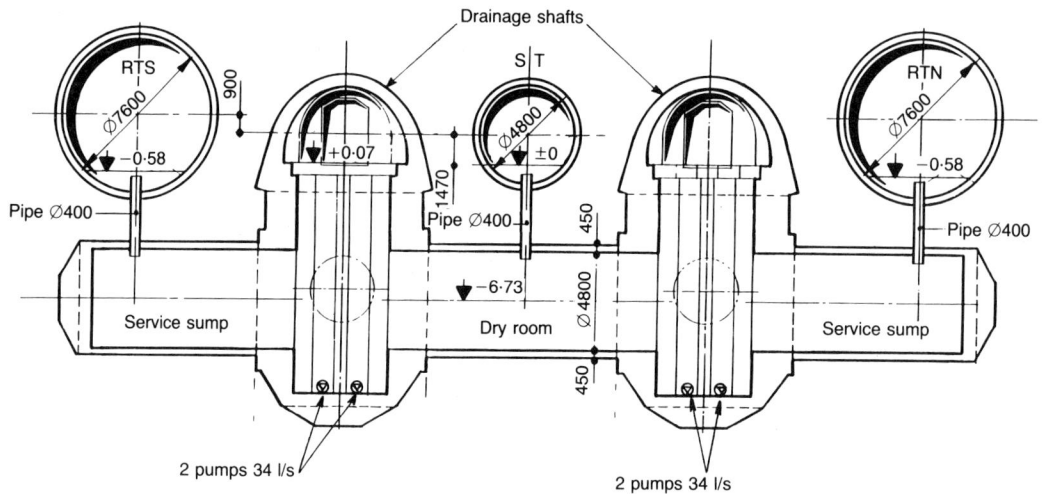

Fig. 9. Section of pumping station on service sump (dimensions in mm, elevations in m above service tunnel)

support of 200 mm shotcrete and lattice frames

(g) shaft 3rd phase to level -12.92 m with temporary support of shotcrete and tie bolts.

Spoil extraction was catered for by wagons waiting at cross-passages.

37.   The extent of the work warranted the introduction and use of a Westfalia Lynx road header with a 110 kW cutting head; 85% of the excavation was carried out by two machines (Fig. 11). A Boehler VBS 14 boring machine fitted with an HB 600H hammer was mounted on a slide fixed to the arm of the road header, the head of which was fitted with a retractable metal arch erector. The remaining 15% of the excavation was carried out by a tracked 3 t mini-shovel fitted with a hydraulic breaker or by means of jack hammers for finishing work (Fig. 12).

38.   The spoil was extracted from the gallery by a PF 1-500 heavy duty conveyor fitted with

lateral drag chains. At the beginning of excavation and for distances of less than 20 m, a conveyor of the EKFO type was used.

39.   The ends of the conveyor emerged in the access adits. In each of these was a 600 mm belt conveyor, 8 m long, delivering to wagons waiting in the service tunnel. A train of six wagons of 10 m³ capacity was despatched to each work station up to the shaft.

40.   While waiting for the concrete lining to be placed, two sorts of support were supplied

(a) rock bolts (Swellex) 1·8 and 2 m long, and 20 cm of shotcrete reinforced by welded mesh (100 × 100)

(b) by means of metal lattice arches (4 × 26 mm) and 20 cm of shotcrete reinforced by welded mesh (100 × 100).

Either type was used depending on the kind of structure.

41.   The shotcrete was delivered in 40 kg bags stored on palettes holding 800 kg.

42.   The Lynx was followed closely by an

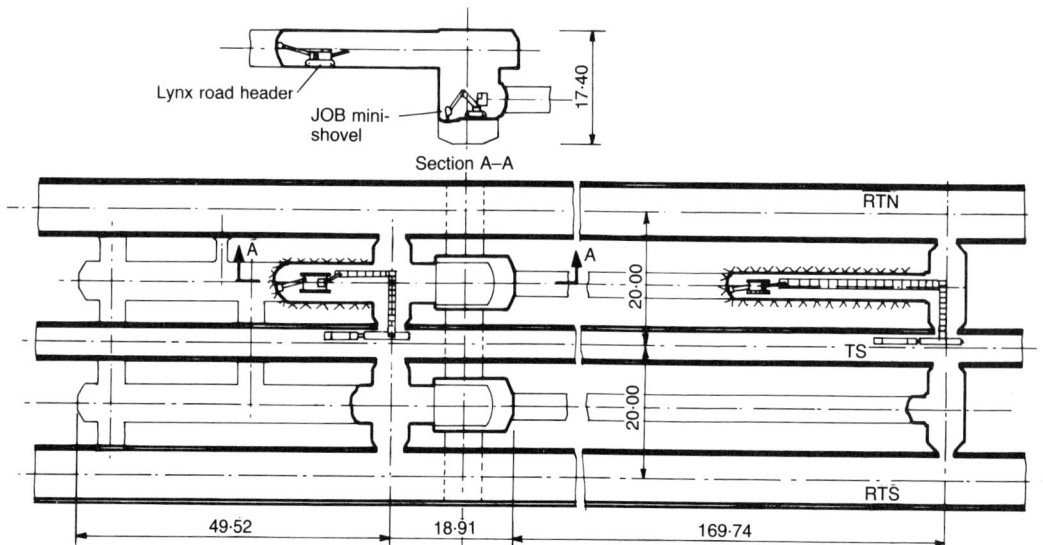

Fig. 10. Excavation of pumping station (dimensions in m)

Aliva 246 machine for spraying concrete, fitted with its own dust filter.

43. The lining was unreinforced concrete. The shuttering was of various types

(a) panels from Tekko Standard Hunnebeck for the invert and the tympanum of the dome
(b) timber formwork for the dome (built in the TML workshop)
(c) Cifa articulated formwork on wheels for the passages and sumps, 4 m long for 4·8 m passages, 6 m long for those of 3·3 m dia.
(d) climbing shuttering of panels 1·2 m high for the shaft.

44. The B 30 TP concrete was mixed at the Beussingue Portal. It was despatched through the marine running tunnels by convoys of three Mulhauser transporters of 10 m³ capacity. Placing was by Putzmeister KOS 1050 piston pumps.

The crossover
45. The communication arrangement between the two running tunnels lies 12·5 km from the Sangatte shaft and is one of two similar major works which divide the route into three reasonably equal parts.

46. The crossover consists of a cylindrical structure of constant section, inside which the two running tunnels are brought together. The connections between the north and south tracks are scissor crossovers. The structure is ovoid and closed at each end by cylindrical tympanums of 20 m radius at the vertical axis.

47. The internal dimensions of the structure are

(a) openings at the springing of the vault: 18·9 m
(b) height along the axis: 12 m
(c) length along the axis: 162 m

and overall measurements

(d) width at base: 26 m
(e) height along the axis: 22·4 m
(f) length along the axis: 170 m.

The shell and the two tympanums are in unreinforced concrete.

48. The crossover lies in the Chalk Marl which at this point is about 37 m thick and which slopes south to north at approximately 30%.

49. As they neared the crossover the running tunnels approached each other until the distance between them was 10·5 m. At this point the service tunnel was diverted to a lower level on the north side.

50. A metal frame floor was fixed below the arch of the tunnels in order to create, under the vault of the crossover cavern, a space which houses the electromechanical equipment for this zone.

51. Reinforced concrete walls starting at the tympanums separated the two tunnels at each end of the scissor crossing. Metal sliding doors filled the gap between these walls.

Construction
52. The structure was excavated open face. Commencement was by way of the service tunnel which was the first of the three tunnels to reach the site. In order to reduce the pressure of demand on the service tunnel, the running tunnels were used to bring in the concrete supplies.

53. Two fixed installations served the site. They served different functions

(a) a workstation dedicated to excavation and spoil removal, situated in the service tunnel near the tympanum on the French side
(b) a workstation in each of the running tunnels dedicated to concrete distribution.

Fig. 11. Excavation of pumping station sumps with the Lynx machine

Fig. 12. Mini-shovel digging in pumping station shaft

54. Ground investigations were carried out from the service tunnel before the cross-passages were driven. When the primary headings were driven they revealed the possible presence of water-bearing fissures or other potential sources of water inflows.

55. Access to all parts of the site was by means of two lateral ascending ramps with a gradient approaching 10%, which allowed all necessary investigative, excavating and spoil machinery access (Figs 13 and 14).

56. The eleven primary headings which made up the shell were constructed in turn, starting with the lowest, on the principle of excavation and immediate concreting (Fig. 15).

57. To keep to schedule, the shell was constructed by three teams simultaneously. The tympanums were built on the same principle. Each segment of the tympanum was excavated and concreted at the same time as the initial heading to which it was connected.

58. Any support of the headings was assured by rock bolts and, if necessary, by shotcrete.

59. After completing the north ramp and the upper cross-passage No. 2, the first team started on the centre heading at the top (Fig. 16). At the same time the south ramp was excavated by the second team as far as the junction with the upper cross-passage. This team then started on the first abutment gallery, south, which was the bottom section of the shell (Fig. 17). The third team proceeded by the same

method to excavate the lower abutment gallery on the other side (Fig. 18).

60. While team 1 was excavating the vault galleries in descending order, teams 2 and 3 were excavating the abutment galleries in ascending order.

61. When the lower headings were being driven, the stresses caused by the partly excavated tunnels were translated into pressures on the lining void. To avoid excessive deformation, the lining was reinforced by HEB 120 reinforcing arches at the rate of two per ring.

62. The core was then excavated in several stages under the protection of the completed shell (Fig. 19). The segments in the running tunnels were dismantled at the same time as the third phase, which consisted of removing the ground between the two tunnels.

63. The invert was constructed after the lining segments of the north and south running tunnels had been dismantled. Phasing of work on the invert allowed a track to be kept free in each tunnel for crossing the site and access beyond the crossover.

64. Excavation was carried out by three teams, each equipped with a Lynx road header having 110 kW of power at the cutting head. The compact design of these machines allowed them to be used in all sectors of the work, in the ramps, access adits, and the primary galleries or the tympanums, without significant dismantling.

65. Removal of spoil was by the same

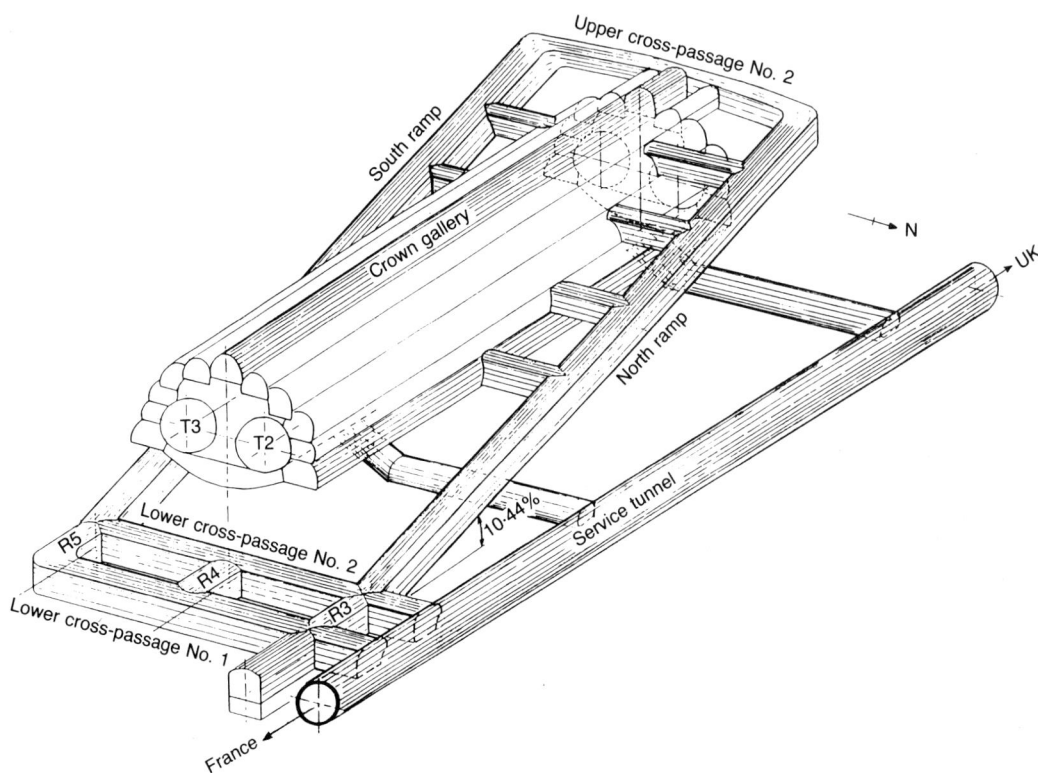

Fig. 13. Crossover galleries and construction passages from service tunnel

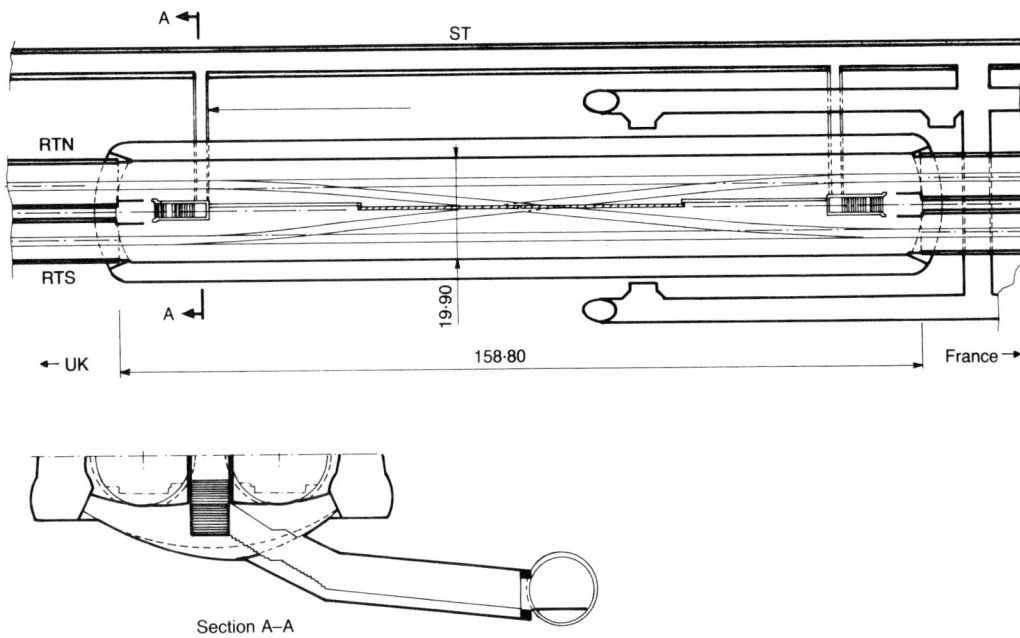

Fig. 14. Plan of
completed crossover
with section of
communication
passage (dimensions
in m)

method as for the pumping station. Spoil tipped
behind the road headers was taken to the
service tunnel by shielded conveyors and conveyor belts operating in the ramps and the construction adits (Fig. 20).

66.   The spoil was then loaded into wagons
of 10 m^3 capacity in convoys of five wagons.
The capacity of extraction in the service tunnel
was 48 m^3/h of ground in place.

67.   The primary galleries were filled with
concrete by successive pours of 11 to 16 m,
depending on circumstances. Three galleries—

C1, 2 and 3 rested against the extrados of the
running tunnel segments. In galleries V1 and
V2 the intrados of the crossover cavern was
shuttered using a hard expanded polystyrene.

68.   The advantage of a mobile concreting
supply was that concrete could be placed in two
different headings.

69.   The concrete was mixed in a batching
plant installed at the head of the Beussingue
cutting. A retarding agent allowed the beginning of the set to be delayed by eight hours.
Concrete was loaded into special mixer lorries

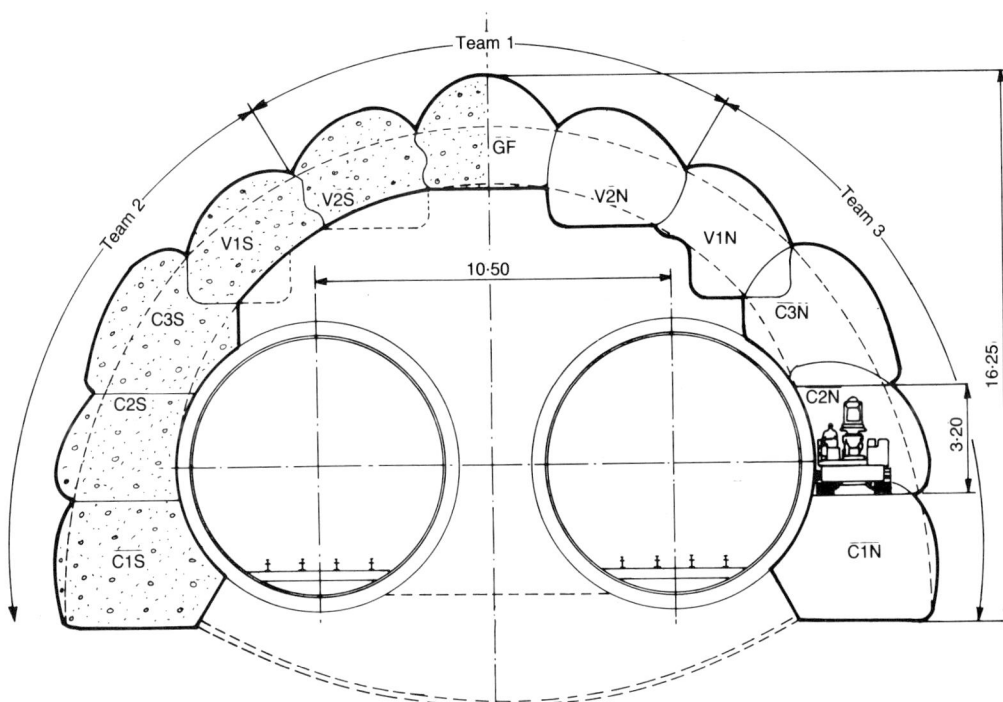

Fig. 15. Works
phasing of the
crossover shell:
excavation (right) and
concreting (left)
(dimensions in m)

Fig. 16. Crossover crown gallery (GF)

Fig. 17. Bottom crossover abutment gallery (C1) showing props on running tunnel lining

and taken to the concreting relay points sited level with the crossover in each running tunnel. At these points were Braima mixer hoppers where the concrete was refined and remixed. It was then transferred by pump to the site. Concrete trains carried 60 m³; the supply capacity was 36 m³/h.

Stages and methods

70.   The support facilities which allowed the primary galleries to be excavated were supplemented by the provision of two loading hoppers which delivered into the running tunnels

(Fig. 21), and two construction adits to the service tunnel. These works were intended for the extraction of spoil from the core excavation (by way of the hoppers) and the invert (by way of the adits).

71.   The top heading immediately under the crown of the crossover cavern was excavated down to the top of the extrados of the running tunnel segments by three Lynx machines at the face, the centre one being 20 m ahead of the other two.

72.   The ground extracted was tipped onto the shielded conveyor and the conveyor belts by way of the south ramp as far as the wagons waiting in the service tunnel. Afterwards, the loaders tipped directly into wagons in the running tunnel through the hoppers.

73.   The next part of the core was excavated to 3·90 m above the level of the temporary tracks, by two road headers working from each end (Fig. 22). The central section was excavated in two stages—after removal of the segments in each of the running tunnels, successively.

74.   Segment removal was subcontracted to STIPS of Algrance.

75.   Dismantling was started as soon as the core had been excavated. A Lynx machine made a cut in the lining from which the top segments were broken down (Fig. 23). This was completed after the upper part of the temporary arches had been removed and the temporary tracks replaced by fill. The invert segments were then removed mechanically by traction. In

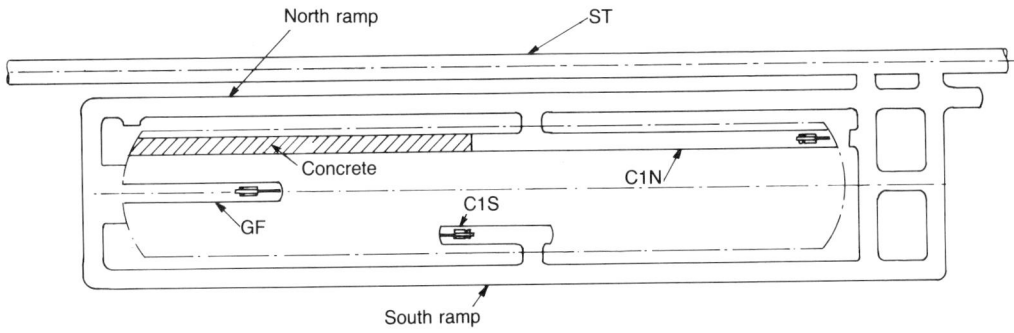

Fig. 18. Plan of first three crossover galleries (C1N, C1S and GF) being excavated simultaneously

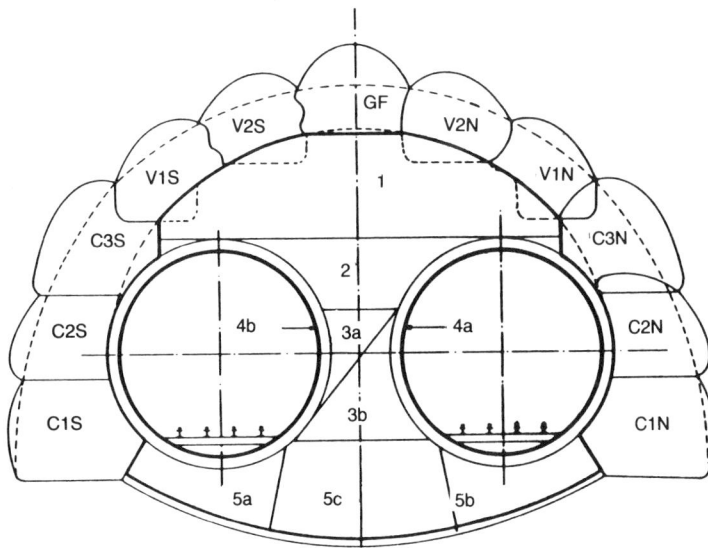

Fig. 19. Works phasing of the crossover core and floor

Fig. 20. Construction passage opening with shielded conveyor

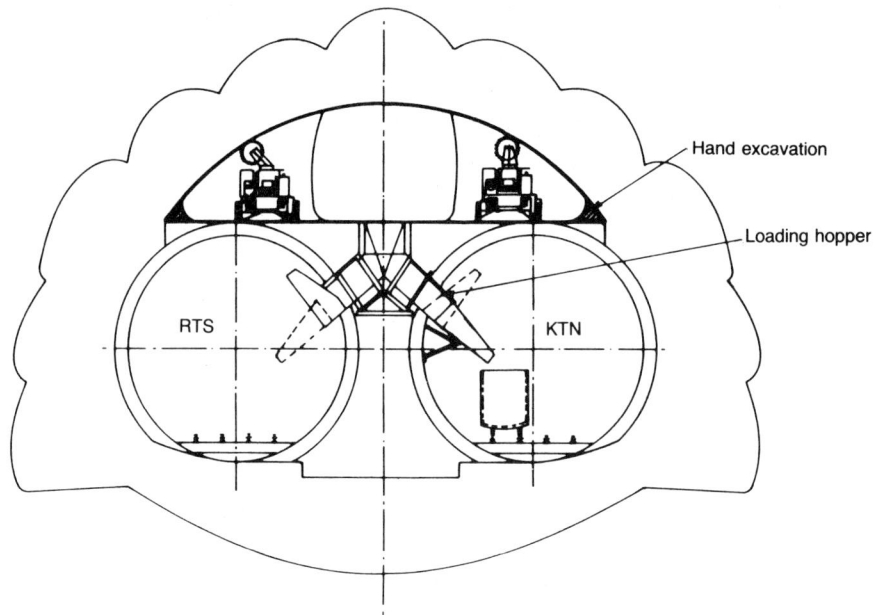

Fig. 21. Phase 1 excavation of crossover core

the tunnel, the segments were broken down by a rock breaker which moved on a track on top of the fill.

76.   The debris was taken to the tipper of the land running tunnel situated at the terminal and later disposed of at the Fond-Pignon discharge site.

77.   Removal of the 150 rings from the two tunnels took six weeks.

78.   The tympanums were completed from the top down, in stages, as the core and invert were removed.

79.   The rock face in the upper half of the cavern was secured by rock bolts in staggered spacing every 1·5 m. The work was done from scaffolding of concrete reinforcing bars, later left in place in the concrete. The shuttering was of timber, kept in place by rods fixed to the heads of the rock bolts. The concrete was placed by pump.

80.   The invert was excavated by two road headers, starting from the temporary cross-passages to the service tunnel (Fig. 24).

81.   This work was carried out in three

Fig. 22 (above). Excavating crossover core (phase 2)

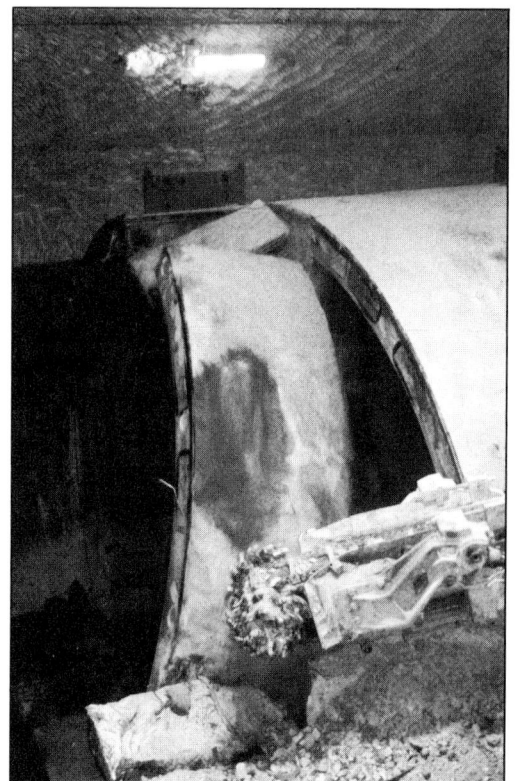

Fig. 23 (right). Destruction and removal of running tunnel lining in crossover (phase 3b)

Fig. 24. Excavation of crossover floors (phase 5)

stages—the sections under the running tunnels in turn and then the central bank. Spoil was taken to the passage by way of a construction adit through the central bank.

82.   Concreting of the invert and the neighbouring tympanums was effected as soon as the road headers moved forward to the next section.

Proc. Instn
Civ. Engrs
Civ. Engng,
Channel Tunnel,
Part 3:
French Section,
1994, 76–81

Paper 10502

Tunnels—safety

H. Barthes, A. Bordas, D. Bouillot, M. Buzon, Ph. Dumont, J. Fermin,
J.-C. Landry, J.-P. Larive, L. Leblond, J.-J. Morlot, L. Szypura,
Ph. Vandebrouck and B. Vielliard

■ **Safety measures at the Channel Tunnel work sites were exceptional. This Paper shows how and why safety measures were provided from the outset and discusses the top-downwards organization of safety and the care given to training. It describes the many intervention levels in addition to the organization and operation of the central control post and safety measures specific to each work site.**

Health and safety was an integral part of the design of the Channel Tunnel project. A specialist security department joined in preparing technical studies before plans were agreed or calls for tender had been issued for the supply of materials or sub-contracts. At the same time contacts were made with the Labour Inspector, the regional sickness benefit office (CRAM) and the Building and Public Works Industry Safety Department (OPPBTP).

2. The safety structure catered for the prevention of risks by training and raising the safety awareness of personnel at all levels. Some 500 senior staff participated in a two-day seminar, and the supervisors and operatives all received between one and two days' training. In parallel, the company doctors ensured that the personnel were familiar with the techniques used.

3. The safety policy at the French Channel Tunnel site was above all a deliberate choice by the TML board in the person of the Construction Director, France, who personally chaired the company health and safety committee, assisted by his directors of sub-projects, as well as the inter-company health and safety group. In addition, two special inter-company health and safety committees were chaired by the directors of the tunnels and terminal sub-projects respectively.

4. The industrial concept of a site with strong and stable logistics ensured the safety of the operatives and the works. The development of systematic and repetitive tasks helped to make the project as free as possible from on-site risks and enabled the employment of technicians rather than traditional miners. This industrialized concept meant that the company's commitment to recruit more than 75% of the labour force locally could be met, despite the discrepancy between the qualifications of the applicants and the skill level required by the site.

Training

5. A wide ranging training plan, supported by the state and regional authorities, prepared the labour force for employment in technical work, of which it had no knowledge and no understanding of the technology, the risks involved and even less of the means of dealing with them.

6. Training in safety and rescue procedures was integral to the entry training course for the site. It was continued thanks to the participation of the specialist security service which contributed to development and implementation of the programmes.

7. A method of checking aptitude prior to training was set up for the driving of machinery, lifting equipment and for electricians. Skilled personnel were given qualification cards of varying colours depending on their category.

8. Training in safety matters was an everyday matter at the various sites on the project thanks to the communications policy. Video screens were an efficient means of publicizing safety messages, which were also printed in the monthly information bulletins, advice notes and distributed and put on notice boards.

Levels of response

9. On a site where an accident can occur 20 km away under the sea, a dedicated organization with an internal rescue plan needed to be set up. The organization relied on three lines of response and a permanent communications network. The alarm could be raised from any one of 400 radio handsets in use throughout the site through a radio channel dedicated specifically for the purpose, or over the telephone by dialling an internal number from any one of the 800 telephone points on the surface or in the tunnels. The alarm came in to the central safety post which was manned permanently by professional firefighters (Fig. 1).

10. The first line of response at the workstations could offer first aid while waiting for help (almost an hour was needed to cover 16 km). This was the result of an intensive training course in first aid and rescue given to 700 employees (i.e. 1 in 5) and two or three rescue operators per tunnel who had national rescue certificates, including resuscitation skills (350 employees or 1 in 10).

11. The second line of response was an internal team consisting of three rescuers and a duty nursing attendant under the orders of the

second in command of the team of 12 professional firefighters. It had specific equipment—two fire wagons and an ambulance wagon—for rescue in a tunnel. The team was on duty 24 h a day for rapid response to bleepers. The rescuers received 48 h specific training provided by the firemen as well as training for the national certificate in rescue and resuscitation qualifications (150 employees or 1 in 25). They also practised on two half days every month.

12. The third line of response involved external resources such as the national fire brigade.

13. In addition the internal rescue plan provided for managers to be available in rotation by radio telephone for a week's standby duty and to be present within 30 min on site. This required the managers of rescue services, equipment, operations and electricity to react effectively in the case of accident.

14. Involving the employees with the objectives of the company through induction, training, communication (internal newsletter and cable video) and motivation contributed to achieving the best safety record in underground workings in France on this site of 4000 people—of whom half were recruited in 1989 alone. Incidents were half the national average. A spirit of accident prevention reigned throughout the operation and was an integral part of every activity at every level.

Central control post

15. Commissioned in September 1989, the central control post was installed at the head of the shaft on the second floor of the shaft annex building. Its staff supervised and managed the main networks of the site—the electricity distribution system, transport and drainage in the tunnels, and the emergency services (Fig. 2). They received information via the telephone, radio telephone and electrical or computer systems.

16. The post had four functions: central technical management of the electricity supply, central traffic control, drainage and central emergency services. The personnel grew as the site developed. By mid 1991 the post employed 36 people under the direction of a head of department to whom five shift supervisors, a supervisor responsible for the electricity supply and 29 operators reported. The firefighters at the central rescue point did not report directly to the central control post but to the head of security. However, as they worked in the same building they were in constant communication with the central control post and their counterparts.

17. The post gave cover round the clock, providing four operators directing tunnel traffic, one managing the electricity supply and the drainage, one shift supervisor in the

morning and afternoon or one team leader at night, and a duty firefighter at the central emergency post.

Safety throughout the site

18. Safety measures connected with methods of work and the various types of mechanical equipment used in the tunnels were addressed under specific headings: electricity, pumping, ventilation, movement of people and equipment, signalling and communication.

Electricity

19. The electrical installation at the site was totally monitored by the central technical control post. Initially connected to 2000 control or check points in December 1989, there were connections to 4500 points at normal operating capacity and up to 5500 power points at the height of demand from the excavation sites.

Fig. 1. Fire and medical personnel in the central safety post

Fig. 2. Monitoring networks in the central control post

The total capacity of the installation was 8000 connections.

20.   Devised by TML and Merlin-Gerin, the system was constantly adapted by its operators as the site developed. It employed a significant amount of state-of-the-art equipment including

(a)   two Digital PDP 1173 multi-task computers providing remote control and operation of points on site, the machines were identical and each was able to cover for down time of the other without loss of data
(b)   four automatic PB 400 Merlin-Gerin processors for preparing instructions and decoding the control messages of information sent from or received at the shaft, the electricity stations (90 kV and 20 kV) and the generators
(c)   three LAC Compex transmission networks to ensure liaison with the 50 processors in the tunnels and at the surface
(d)   three supply stabilizers to ensure that technical control, administration, the radio and the central internal telephone system were functioning properly, irrespective of fluctuations in the electricity supply (from EdF or generators)
(e)   a console comprising five 19 in screens, four keyboards, three printers, a radio, a direct telephone link with the site and two telephones connected directly to the 20 kV electricity substation and the 90 kV transformer substation.

21.   The working methods and materials used ensured the highest possible safety standards. Non-flammable cabling was routinely used, particularly in sensitive areas—shaft, marshalling chambers, TBMs and substations in the tunnels—which were all fitted with halogen-free non-flammable cables.

22.   Earthing during excavation of the tunnels was ensured by a grid network to which all installations were connected to ensure equipotentiality.

23.   The substations were fitted with a system of fire detection with automatic extinction by halon gas. In addition, a temperature and smoke detection system was placed near each set of medium or low voltage switch gear.

24.   The cables, laid in lengths of 250 m, were connected by waterproof connectors. These in turn were insulated in a thermo-retractable sheath or inserted into a connecting box which was locked.

25.   All the electrical equipment installed in the marine tunnels was waterproof to IP 679 levels of protection (up to 1 h immersion under 1 m of water) or to IP 689 for the advanced drainage stations (immersion for 48 h under 10 m of water). The 20 kV transformers were either protected by nitrogen or insulated with Ugilec-T. The lighting transformers had a minimal filling of oil and were waterproof.

26.   The catenaries were fed by conductor bars in insulating sheath (IP 235). They were disconnected in working areas and where there was to be plant movement (the locomotives then ran on batteries).

27.   Staff at the central technical control post (part of the central control post) supervised all the electrical installations.

Drainage

28.   Monitoring of the drainage by the control centre was by means of

(a)   one synoptic console showing all the pumping installations in the marine tunnels—this gave an immediate picture of each of the pumps, whether in operation, stopped, faulty, or on stand-by, of the level of water in the gravity sumps, as well as rates of flow in the normal and stand-by drainage systems
(b)   five PB 400 processors connected to the control units of the drainage in the tunnels
(c)   five printers which gave a record of conditions and incidents.

Part of the drainage system was connected to the electrical monitoring system which immediately indicated any pumping problems. Every precaution was taken to ensure that the drainage systems functioned continuously.

29.   The electricity supply was provided by a priority network which was backed up by the stand-by generators. All the circuits were duplicated in each marine tunnel. The relay stations every kilometre were fed by two cables which formed a ring main. The supply to the drainage system at the face was backed up by the emergency circuit of the TBM.

30.   Several independent pumping mains were installed in the tunnels. In case of failure of the permanent drainage, extraction was ensured by the emergency system by pumping out the cofferdam at the lowest point after overflow of the upstream cofferdams. The existence of a pumping main of 150 l/s, in addition to that of 350 l/s, ensured a back-up for the emergency drainage. The pumps are duplicated at the low points of each emergency main, so if one pump stopped, its repair or replacement would not interrupt the functioning of a main.

31.   In the shaft, in case of failure of the normal drainage system, extraction was ensured by one of the two emergency mains; a replacement pump covered for the possible failure of one of the three emergency pumps. In addition, if necessary, the buffer sumps could feed to the eight pumps of the spoil extraction system, which could each handle 175–200 m³/h. The capacity of the emergency system in the shaft was therefore increased by between 400 and 450 l/s.

Ventilation

32. The electricity supply for the ventilation system was provided from the backed-up network. In the tunnels the fans were also fed by a ring main to ensure continuity in case of a cable break. The fans or groups of fans had reserve equipment installed in the suction line. It was held on stand-by in the blower line.

33. At the central control post, the central technical control post supervised the operating condition of the fans and their output in real time. The gas detectors were also connected to the central technical control post.

34. The working environment was continuously controlled

(a) at the TBM (with local alarm) by detectors of oxygen, carbon monoxide, carbon dioxide and hydrogen

(b) in the tunnels (reporting to the central control post) by detectors of hydrogen (at the high points and in the crossover), smoke, carbon monoxide, nitrous oxide and rises in temperature every kilometre, at each electricity substation.

Moreover, regular sampling checked the amounts of dust and silica. At all times manual measures could be called on to determine the moisture content and investigate explosive gases such as methane (the presence of which was unlikely in the geological conditions).

35. In case of fire, there was a danger of toxic and hot fumes occurring in the site and certain tunnels. In the land tunnels and at the start of their excavation, foul air was discharged directly to the surface to avoid any pollution of neighbouring tunnels or the shaft. While the marine tunnels were being driven, ventilation was halted in the event of any accident, which prevented fumes being driven from the face. To remove them, the extraction fans only were operated until the fumes had completely disappeared.

36. The primary ventilation system was an important safety factor. In case of fire, especially at the face, the primary ventilation system left the two air supply tunnels free of smoke or pollution, allowing rescue services to operate and people to be evacuated.

Movement of personnel

37. Access to the tunnels from the shaft was regulated by a system of tags for identification and transport. In all instances, for any person present in a tunnel beyond the shaft during excavation, there was a corresponding tag left on a board at the surface to indicate their identity and the part of the site being visited.

38. In the tunnels, personnel were transported in manriders (Fig. 3). These single, double, triple or quadruple diesel operated vehicles could take 16, 38, 64 or 90 passengers

respectively. Their safety was assured by the following arrangements

(a) while moving, the doors and windows did not allow any part of anybody out and an escape hatch was provided in the roof

(b) the driver's windscreen was of laminated glass and had a windscreen wiper

(c) the doors were fitted with a switch so that the vehicle could not start if they were open

(d) the interior was lit and a radio telephone and fire extinguisher were provided

(e) the manriders, like the battery-operated locomotives for supply trains, had a 'dead man's hand' mechanism on the control lever, a radio telephone and a headlight with a range of 100 m in tunnels

(f) the vehicles were fitted with large safety couplings.

Traffic control

39. Each tunnel was divided into three movement zones.

(a) *The shaft* was operated autonomously to a great extent with manual equipment. In each marshalling chamber an operator handled the vehicles when their arrival was announced by direct line from the control post. The operator controlled the tipping operation (emptying and cleaning the spoil trucks) and transport of the rolling stock in the goods hoist in liaison with the services at the surface, programmed the type and number of wagons to make up each complete train, and requested their return to the tunnel from the operator at the central control post.

(b) *The traffic zone* was administered from the central control post by the traffic control

Fig. 3. Personnel travelling in manriders equipped with APEVA respirators

centre. The movement of trains was controlled section by section; each convoy automatically opened its entry to the next kilometre.

(c) *Traffic at the face*, i.e. in the area behind the TBM and all the work sites for special works, was regulated by a despatcher on the last wagon, assisted by the operator in the central control post.

40. The central traffic control post, in control of all traffic movements in the tunnels, employed four operators round the clock—one for each of the marine tunnels and one for the three land tunnels. Each operator was responsible for the part of a tunnel between the marshalling chamber at the bottom of the shaft and the last wagon of the TBM back-up. The operator could control the traffic automatically. The operator's job was to prepare the traffic plan for a mobile unit by taking into account movements of other convoys and trains; to check that the automatic controls were working and that the driver of the vehicle, communicating by radio, was following instructions. Alternatively the operator could take direct control of the operations by radio. This happened if the adjacent zone was occupied or if there was a fault in the automatic system.

41. The traffic control post had a screen displaying all the tunnels, the shaft and the TBMs during the excavation period (Fig. 4). It had four Merlin-Gerin P400 processors preparing instruction data and analysing incoming information, data transmission networks, electrical supply stabilizers, a desk with three 16 in screens displaying moving traffic, three keyboards, a control desk for preparing itineraries for automatic application or visual control, and the means to contact drivers by radio and telephone. Each controller had a site telephone and two telephones giving direct contact with each end of the zone being controlled.

Fig. 4. Traffic control mimic diagram in the central control post

42. Traffic control was by means of the signalling system comprising

(a) *traffic lights*—red (stop), green (way clear) and violet (visual operation; permission to pass to next section depends on the operator in the central control post)

(b) *convoy detectors* between the rails signalled the passage of a vehicle and distinguished priority of manriders using a magnet beneath them

(c) *points detectors* controlled the points, and therefore the route—at each set of points a luminous orange arrow indicated the position of the points, left or right

(d) *motorized points mechanisms* with an irreversibility system after a change of position

(e) *programmable processors* allowed automatic control of traffic—in the event of equipment failure, the signal showed red as a fail-safe.

Communications

43. There were alarm systems at all crucial points in the installation which were permanently monitored by the central control post operators. The alarm could be raised by radio, by telephone, by warning lamps, or by anomalies on screen or print-outs in the control post. If there was a problem, the control post would take action in accordance with prearranged procedures.

(a) *Power supplies*. The control post would seek the fault (at the main 90 kV station, the secondary 20 kV station or specific substations) and seek relevant assistance and isolate the incident.

(b) *Drainage*. The control post would contact the staff involved, reset the equipment or start up the emergency system. In the event of prolonged breakdown of any emergency system the special works were halted and if problems were met on all stand-by networks, the TBMs were also stopped.

(c) *Ventilation*. The first line response would be called up, the faulty fans restarted if necessary, the person in charge in the tunnel concerned would be warned and could then close down the ventilation in the event of fire.

(d) *Pollution in the tunnel*. Computer screens were watched to detect various gases. In the event of a gas escape the duty firefighter and, if necessary, the fire officer and the rescue services manager would be warned. The operator could call for work in the tunnel to be stopped or for the tunnel to be evacuated.

44. The central emergency post intervened each time people or materials were at risk. In case of injury they contacted the site medical

service, gathered information about the condition of the injured party, guided the first aiders to the spot and contacted the services needed (manriders, public ambulance service, police, etc.). In case of fire, on-site or external fire services could be called, depending on the seriousness of the incident.

45. In the event of technical problems in a tunnel, the central emergency post was alerted by the traffic controllers or by the general technical controllers. If people were injured, the first service involved had to contact the emergency services direct.

46. Finally, in order to ensure efficient control from the central post the duty manager had two workstations.

(a) *A console in the central control post with eight telephones*. One telephone was reserved for incoming calls, another for outgoing external ones. Three network lines (one a TML switchboard line, one through the local public telephone system, and one international line) connected directly with the UK control post. Three lines connected directly with the TBMs.

(b) *An office attached to the central control post*. Access to this was unrestricted, and it contained three telephones—two reserved for outgoing and incoming calls and one to the local public switchboard in order to provide a link with the outside world in case of an incident affecting the site's own telephone system. A personal computer was available for the duty manager to prepare a daily report of activity in the control post.

Proc. Instn
Civ. Engrs
Civ. Engng,
Channel Tunnel,
Part 3:
French Section
1994, 82–87

Paper 10504

Tunnels—the terminal

H. Barthes, A. Bordas, D. Bouillot, M. Buzon, Ph. Dumont, J. Fermin, J.-C. Landry, J.-P. Larive, L. Leblond, J.-J. Morlot, L. Szypura, Ph. Vandebrouck and B. Vielliard

■ **The terminal supports all the facilities required by the tunnel user, and those necessary for the management, the maintenance and safety of the transport dimensions and functions, it is comparable to a major international airport. After having defined the functions of the infrastructure, this Paper looks into the studies that preceded its completion. It also discusses the works carried out on a soil with limited bearing capacity. A section is devoted to the civil engineering structures and the installation of the railways.**

Fig. 1. The Channel Tunnel within the European rail (continuous line) and road (broken line) networks

The construction of the French terminal does not initially reveal any original or unusual construction techniques. Nor is the site especially large. The real difficulties lie in the initial concept of the system and the interface between the design, construction and commissioning.

2. The junction of the Eurotunnel transport system with the European road and rail networks on both sides of the Channel (Fig. 1), the terminal not only provides all the necessary services for travellers but also provides for the operation, management, maintenance and safety of the transport system.

The functions of the terminal

3. The terminal's primary function is to funnel the traffic from a three-lane motorway (3450 vehicles per hour) on to a system of shuttles having first dealt with the formalities of tolls, immigration and customs.

4. The second involves getting the shuttles into the tunnel so that they complete the journey from platform to platform within 35 min, while at the same time ensuring the passage of conventional trains and the high speed trains (TGV) from the French railways (SNCF) at the rate of 20 trains and shuttles per hour. The third function is that of a maintenance depot for Eurotunnel's rolling stock.

Engineering design

5. The design of the terminals, while taking into account the specific character of each site and the need to integrate with the environment, had to give the user a sense of comfort and continuity as he or she leaves the road network, is taken into charge by Eurotunnel, and is released on the road to his or her destination.

6. The Coquelles Terminal forms part of the fixed link, which consists of

(a) the civil engineering works of the tunnels and the terminals in France and the UK
(b) trackwork, signalling, electricity supply, communication networks and tunnel equipment
(c) supply of specialized rolling stock and maintenance facilities
(d) operations, border formalities and safety facilities.

Consequently these activities had to be interfaced with the two terminals at the two extremities of the 50 km project.

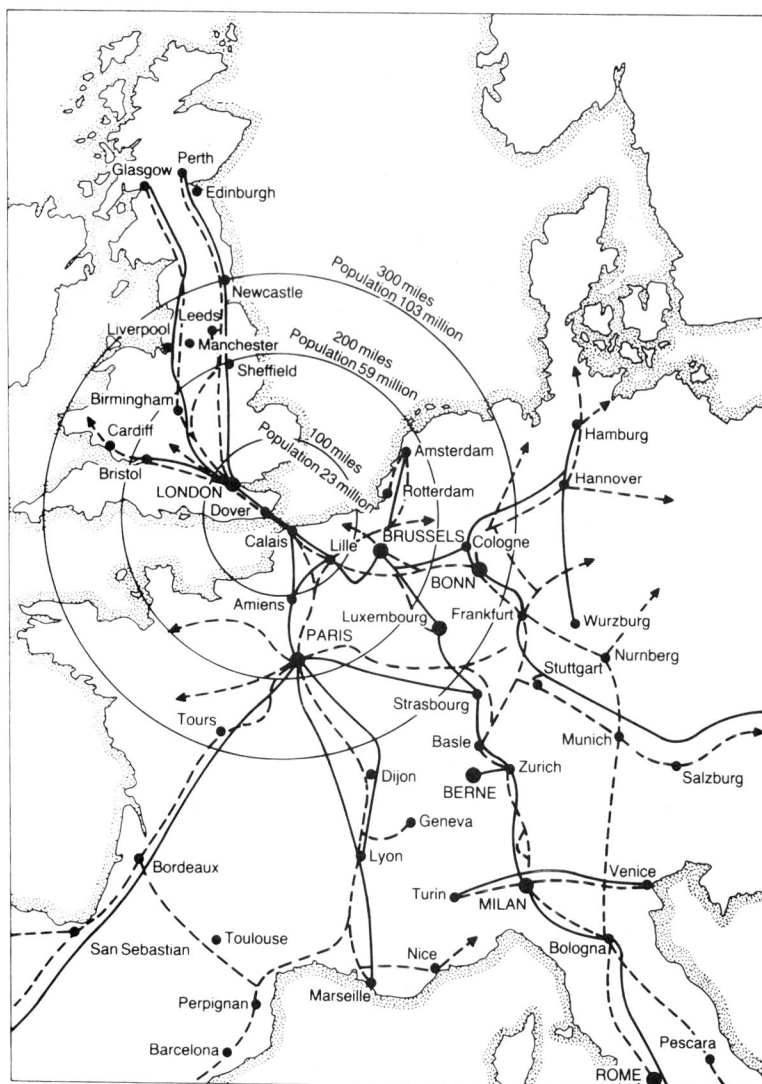

The design process

7. As soon as the contract with Eurotunnel was signed, work began on the development studies, consisting of a general review of the project. In fact, when the call for tender went out and in the short time allowed to prepare and submit an offer, the interested companies had no time to undertake any fundamental studies. These development studies soon exposed the problems inherent in rolling stock and fixed equipment designs which had repercussions on the civil engineering works.

The performance contract

8. The performance contract was defined in general terms in Schedule 2 of the construction contract. It included the following constraints.

(*a*) Eurotunnel was obliged to comply with the clauses of the Concession Agreement which was granted by the French and British Governments and had, consequently, to take into account the requirements of the intergovernmental commission, the decisions of which were final. As a result, the performance was affected, even if these effects had only a limited impact on the terminal itself.

(*b*) Eurotunnel was faced with unpredictable external decisions such as the French motorway plan, which involved a change to the local road network, or the introduction of the EC *acte unique* after the contract was signed—the exact consequences of which have yet to be calculated.

In such circumstances, the role of the engineers on the French terminal was particularly difficult as they also had to meet the pressing demands from the ongoing construction work.

Traffic considerations

9. The design of the infrastructure of the terminal was based on the assessed throughput in one direction for the thirtieth peak hour traffic levels for the year 2003 (given in various categories of road vehicles). The traffic throughput figures in one direction for tourist and freight traffic are given in Table 1. The possibility of extending the installation as the traffic increases to saturation point was taken

Table 1. Thirtieth peak hour one-way traffic in 2003 (vehicles per hour)

Vehicles	Thirtieth peak tourist traffic hour	Thirtieth peak freight traffic hour
Cars and light vehicles	785	260
Coaches	74	25
Heavy goods vehicles	55	130

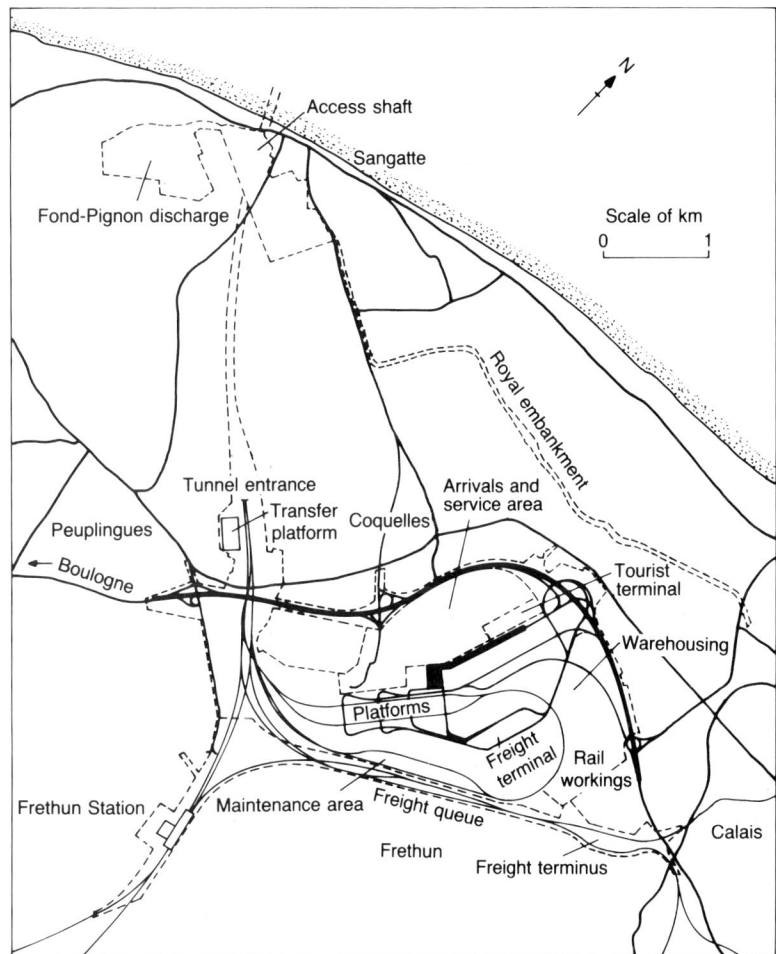

Fig. 2. Location of the terminal

into account in the conception of the general layout design. The terminals were also arranged to cater for significant seasonal, weekly and daily variations in road traffic.

Location and site layout

10. The Coquelles Terminal occupies a land area of about 650 ha of which 450 are taken up by the installations directly connected with the fixed link (Fig. 2). It covers four parishes: Calais, Coquelles, Frethun and Peuplingues. It is bounded to the north by the route nationale 1 from Calais to Boulogne; to the west by the Paris–London SNCF railway line and the Beussingue Portal; to the south by the Paris–Calais railway line and to the east by the River Neuve canal and the future west access for Calais (Fig. 3).

11. Generally, the work was divided as follows

(*a*) ground consolidation, general earthworks and drainage: 40%

(*b*) roads, car parks and other traffic areas: 10%

(*c*) civil engineering, road and miscellaneous structures: 30%

(*d*) buildings: 20%.

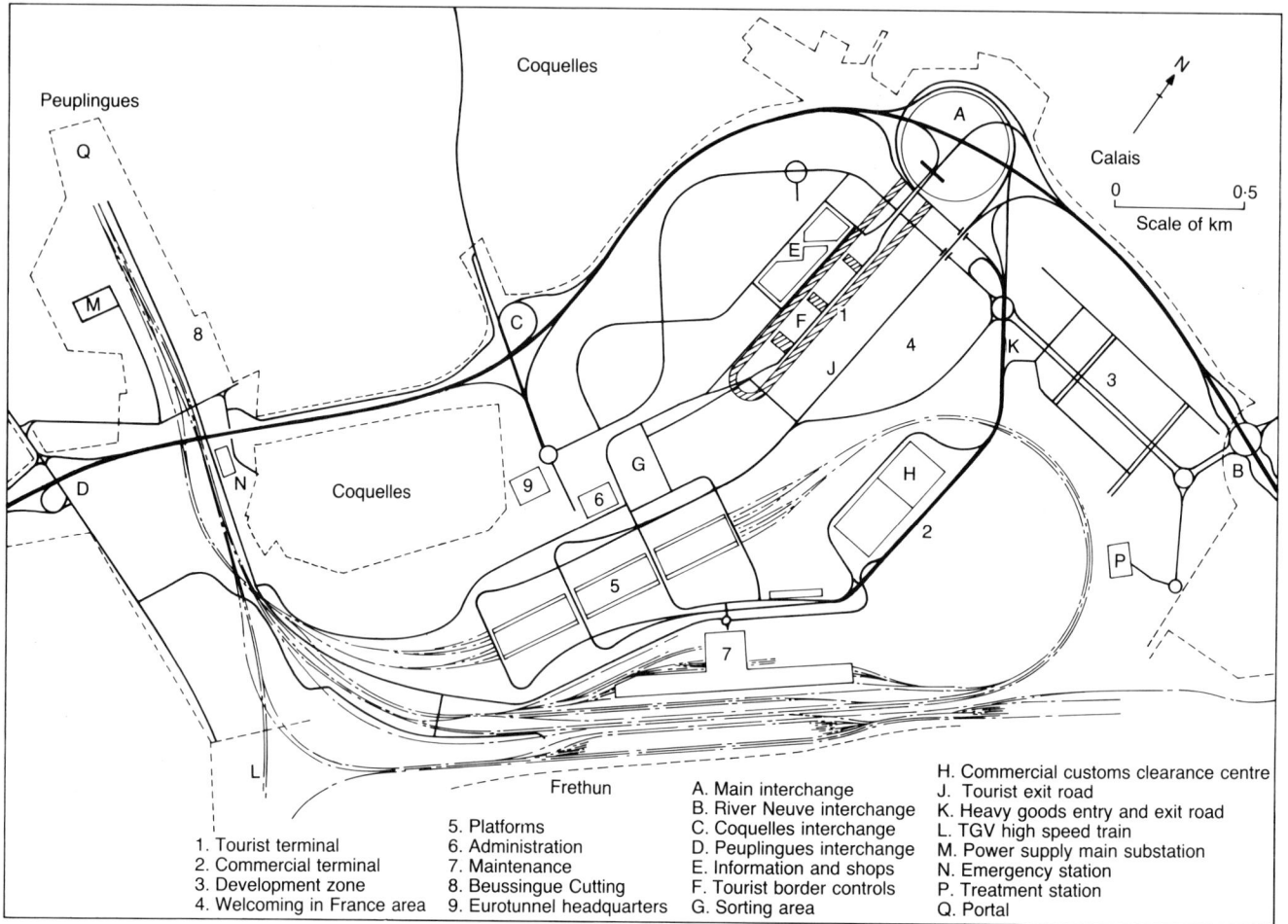

Fig. 3. Terminal layout

1. Tourist terminal
2. Commercial terminal
3. Development zone
4. Welcoming in France area
5. Platforms
6. Administration
7. Maintenance
8. Beussingue Cutting
9. Eurotunnel headquarters

A. Main interchange
B. River Neuve interchange
C. Coquelles interchange
D. Peuplingues interchange
E. Information and shops
F. Tourist border controls
G. Sorting area
H. Commercial customs clearance centre
J. Tourist exit road
K. Heavy goods entry and exit road
L. TGV high speed train
M. Power supply main substation
N. Emergency station
P. Treatment station
Q. Portal

These are comparable with work on a major international airport.

Geological limitations

12. Geotechnical constraints, in particular the compressible peaty ground, influenced the configuration of the French site and the positioning of the rail loop between the SNCF railway tracks and the Calais–Boulogne motorway. This led to a clear separation between the terminal installations and the main European road and rail links.

13. The choice of the trackwork layout along with the high water-table and the nature of the ground influenced the general layout of the terminal. The aim was to limit the amount of imported fill (rare in this region) and the amount of ground compaction. This resulted in the following layout

(a) the tourist terminal situated between the new motorway and the rail loop on sound ground, slightly above the rest of the site

(b) the commercial terminal for heavy goods vehicles situated in the rail loop on peaty ground

(c) the embarkation and disembarkation platforms situated alongside the tourist ter-

minal, as far as possible outside the peat zone

(d) the rolling stock maintenance facilities situated to the south on peaty ground between the platforms and the SNCF railway lines.

The terminal and its users

14. The design of the terminals is based on the principle of maximum separation between tourist traffic and heavy goods traffic from the main entrance road to the platforms. The problem of directing users from the national road network, despite the constraints specific to each site, is the same at both the UK and French terminals. The French situation is described here.

15. From the road interchange at Fort Nieulay—a strong architectural feature with a 400 m diameter lake—the installations lie along a single axis. The alignment was considered carefully to ensure that users could view all the installations as soon as they entered the terminal area. They then pass:

(a) the entrance barrier coupled with the tolls (twelve lanes equipped with double booths serving two adjacent lanes)

(b) the border and security checks before leaving the country

(c) the sorting area where vehicles are separated and organized according to type and allocated into four groups corresponding to the capacity of the four types of single or double deck shuttles.

Finally they reach the platforms.

16. Allowing for the position of the tourist and commercial terminals along with the trackwork layout, the platforms consist of

(a) two embarkation overbridges accessible on one side to tourist vehicles and on the other to heavy goods vehicles

(b) two disembarkation overbridges with exits to the north-west for tourists and to the south-east for heavy goods vehicles.

The four overbridges are situated at right angles to the parallel platforms and are of the same length.

17. This layout allows the tourist vehicles to be separated completely from the heavy goods vehicles both at the entrance and the exits while guaranteeing a great deal of flexibility.

18. The organization of border controls and security is based on

(a) separation of tourists from heavy goods vehicles

(b) separation of French and British controls into separate areas

(c) separation of vehicles subject to lengthy examination from general traffic to avoid bottlenecks

(d) unimpeded exit from the Eurotunnel system.

As there is unimpeded exit, the user may, after disembarking from the shuttles, proceed directly to his or her destination without any further checks or delays.

19. Temporary vehicle waiting areas capable of holding up to three hours' traffic are positioned along the users' route to cater for super peaks or traffic accumulation resulting from an incident.

The railway layout

20. After much work was carried out to optimize the trackwork layout from 1986, the loop solution was retained because it gave the best flexibility. The layout was the result of

(a) the alignment of the tunnels between Sangatte and Beussingue

(b) the geometry of the loop

(c) the high-speed approach of the shuttles

(d) the requirement to connect to the national railway network and the maintenance area

(e) construction of platforms and tracks as far as possible on solid ground.

In addition, the topography and the track alignment also necessitated a rail overbridge.

Architecture and the environment

21. Another relationship between the terminal and the user centres on the perception of the architecture and landscaping. The general architectural design sought to achieve harmonious integration with the local and regional development plans, a coherent plan (despite the large number of functions to be provided for and future expansion of the system), integration with the natural landscape, and the manifestation of Eurotunnel's corporate image throughout the system. On this basis the French architects, Paul Andreu and Pierre-Michel Delpeuch (Paris Airports consultants), collaborated with TML to draw up the architectural plans for Coquelles Terminal.

Safety

22. The workings of the installations of the fixed link had to meet well-defined safety conditions both inside the tunnel and on the terminals. The design requirements in this sensitive area were mainly the prevention of risk from natural and man-made hazards through a central alarm system, and emergency plans to deal with major incidents (e.g. fire, explosions).

23. To meet these criteria the following facilities were constructed at the terminal site

(a) an underground fire water main fed from two independent supplies

(b) a rescue centre housing the rescue staff, and a garage for the service tunnel transport rescue vehicles

(c) an emergency siding track to receive trains in difficulty

(d) access for external and internal emergency services

(e) an emergency train (held in the maintenance area) to provide heavy emergency equipment.

Works carried out

24. Extending over 650 ha, the construction works comprised: about 11×10^6 m^3 of earthworks (4.5×10^6 m^3 chalky or silty excavated material, 3×10^6 m^3 imported gravel for fill, sand filter beds and formation level structures, and 3.5×10^6 m^3 surcharge material for ground consolidation (Fig. 4)); a drainage network comprising 180 km of open channels, ditches, drains and pipes; 10^6 m^2 of paved areas; structural works (58 000 m^2 of bridges, 4300 m^2 of platforms); 60 buildings with a surface area of 31 000 m^2 (Fig. 5); and about 50 km of track and 90 turn-outs and points. On the technical side two aspects in particular deserve

Fig. 4. Fleet of earthworks machines in the Beussingue Cutting

attention—ground consolidation and the drainage system.

Drainage system

25. The question was how to collect and remove water from a flat area where the subsoil was largely peat and clay with a water-table close to the surface. There were no low points, so classical methods could not be used. An original solution was found based on the creation of raised stormwater retention lagoons; channelling of the water by gravity to the foot of the stormwater retention lagoons; and raising the

Fig. 5. Terminal maintenance building

water into the stormwater retention lagoons by pumping. Water flows by gravity from the first lagoon to the foot of the following lagoon and so on through a system of channels until the water reaches the west of Calais where it is released into the port at low tide.

26. The drainage system comprises five raised stormwater retention lagoons covering almost 100 000 m²; six pumping stations with a total capacity of 85 000 m³/h arranged at the foot of each lagoon; 10 km of channels and drains connecting to the Pierettes Canal and the River Neuve. The underground and surface waters from the Beussingue Portal are collected close to the Portal and raised by pump to the crest of a slope from where the water flows by gravity to the foot of the first lagoon. The total capacity of the lagoons is 250 000 m³.

Structural works

27. The quantity of bridge work varied little from the initial 58 000 m² envisaged in 1986, but there were significant modifications to the structures as the design progressed. Examples are the change of the Boulogne-sur-Mer road connection from a coastal by-pass to a motorway, the reduction in the number of access ramps to the platforms from ten to eight, and, most importantly, the architect's involvement in the design of the structural works.

28. In the initial design most of the bridges were planned as isostatic deck and beam structures. Studies carried out by the terminal's own engineering department (BETER) and cost estimates of different types of decks, confirmed that beam bridges were the most economic solution and had two important advantages for the terminal

(a) reduction of the propping which would have been necessary with any other solution and consequent elimination of the need for any ground consolidation

(b) assembly in independent spans which made it possible to organize erection in phase with the foundation works, which in turn made it possible, if necessary, to provide access beneath the spans for other work such as drainage.

29. The remaining structures were to have been slab bridges or boxed bridges built with supported decks poured in situ. However, in the event, most of the road bridges—particularly the road interchange of Fort Nieulay—were built with continuous concrete slabs poured in situ.

30. The four overbridges to the platforms are spaced over 1 km. They straddle the group of tracks alongside the platforms to which access is by means of ramps (Fig. 6). The overbridges consist of 20 bays of 10 m–20 m span. The decks are isostatic frame structures placed on trimming beams supported on 1·6 m dia-

meter columns. The ramps were constructed on the same principle. The foundations are piled. The whole work covers an area of 36 000 m². The piers and the trimming beams were poured in situ with metal shuttering which was reused between 40 and 100 times.

31. The deck was then constructed by placing the beams on the trimming beam, placing precast slabs (prédalles) between the beams and overhanging the side beams, installing the deck reinforcement, concreting several spans at a time and then placing the moulded parapet units. The beams and prédalles were manufactured on site and the moulded parapet units were supplied by a factory in Calais. The precasting plant was within the site installation area.

32. These works amounted to 87% of the total. Other works included mainly boxed or slab bridges, which are not of any special interest except for the angled rail overbridge which forms a figure eight trackwork configuration.

33. The platforms are significant because of their great length (850 m) and number. In section they are U-shaped and were not difficult to construct. Being of uniform design, they allowed the shuttering to be reused several times. The only difficulty with the platforms lay in the fact that they were built in an area made complex by the amount of interference from other activities, such as surcharge material movements for ground consolidation, drainage works, cable laying, trackwork and electrical catenary feeders for the railway locomotives.

34. *Fort Nieulay interchange* Orientated on the axis of an old military fort, the interchange evolved because of the architect's wish to create a major architectural feature at the access to the tunnel. The main feature is a large Y-shape viaduct made up of 36 spans crossing the 400 m diameter lake, and eight small structures of three or four spans. It was built as a slab structure by the Rateau, Borie, Nordpac group, and covers 12 000 m². The foundations are piled and the flared columns are supported on the pile caps. Because of the long spans, the deck had to be of prestressed concrete in order to reduce the number of support points and to allow speedy removal of shuttering. This technique reduced costs as well as time.

Conclusion

35. The difficulties in the construction of Coquelles Terminal did not lie in the complexity of the construction techniques, but rather in its integration within the much greater task of designing, building and commissioning the entire fixed-link transport system in just seven years. The TGV high-speed train service, for example, took ten years from initial design to completion of the first (south-east) line.

Fig. 6. An overbridge with platform access ramps

*Proc. Instn
Civ. Engrs
Civ. Engng,
Channel Tunnel,
Part 3:
French Section,
1994, 88–92*

Paper 10505

Tunnels—transport system and fixed equipment

*H. Barthes, A. Bordas, D. Bouillot, M. Buzon, Ph. Dumont, J. Fermin,
J.-C. Landry, J.-P. Larive, L. Leblond, J.-J. Morlot, L. Szypura,
Ph. Vandebrouck and B. Vielliard*

■ **The Channel Tunnel is a complete transport system. The Paper explains how and why a specific organization was set up, the different aspects of the undertaking to construct the system, its original features and the resources used. In an infrastructure of this type, the fixed equipment in the tunnels as well as at the terminal is of primary importance. This involves the railway tracks, all the electrical networks, the control and communication facilities as well as the ventilation and aerodynamic systems. The Paper also describes tunnel drainage, fire-fighting systems and the organization of utility rooms.**

The transport system

The establishment of the transport system required the setting up of a specific organization. The various aspects of this construction activity, its unique features and the resources used are described in the following paragraphs.

2. Based on studies by the Engineering Group and with their support, the TML Transport System Group had the task of setting up the fixed equipment and the rolling stock for the tunnel transport system. This work was done within the terms of the programme, costs and performance levels set by Eurotunnel and Transmanche Link.

Unique features of the task

3. The transport system is unique in the sense that there were no precedents to call on for a project of this complexity and, although traditional techniques were used, they were subject to extremely high demands of performance, safety, and interaction with conditions within the infrastructure (tunnels) rarely met before.

4. The undertaking was a Franco-British project, with teams from both countries. It was financed by international private capital and owned by a recently established client having no experience. It had to form a link between two well-established transport networks of very different constitution and organization and between which there had hitherto been no practical exchange.

5. The project also had a very tight schedule. Moreover, the fixed equipment for the transport system had to be undertaken at the same time as the tunnel was being driven, and from two points of access, one in France the other in the UK, about 50 km apart.

6. The rolling stock was an entirely new development; its cross-sectional dimensions required special test areas.

7. The technical organization plan was, with only slight modification, taken from the same breakdown of work used by the Engineering Group. This approach ensured that the engineering documents required little restructuring and that the 'primary system dossiers' were simply split by work scopes.

Fixed equipment policy

8. The policy for the fixed equipment was essentially to subcontract. The Engineering Group studies, together with contractual clauses which reproduced all or part of the contract between Eurotunnel and TML, formed the basis of the international calls for tender published in the official journal of the European Community and in sections of the press. The contracts were let after responses to pre-qualification criteria and competitive tendering had been obtained. They were 'performance' contracts in which performance specifications for each of the sub-systems were defined as clearly as possible, the detailed means of achieving them being left up to the 'know how' of the subcontractors. This type of contract is relatively unusual and is akin to design and build contracts.

9. The contracts included the following stages: detailed engineering; supply; the training of the future operator; installation or its supervision; commissioning; period of guarantee.

Installation of fixed equipment

10. Fixed equipment was installed in the terminal areas and in the tunnels. Traditional installation methods were employed for the terminals but the tunnels presented a particular challenge. Effectively, about 50 km of tunnels had to be fitted out from their two extremities. The tunnels include about 500 underground rooms. The civil engineering work continued at the same time as the installation of electro-mechanical equipment.

11. A remarkable temporary transport system was installed in order to cater for the civil engineering needs and, at the same time,

to ensure transport for the electro-mechanical equipment and its installation (Fig. 1). The installation procedure also had to be adaptable to any unexpected changes in tunnelling plans.

12.   Access to different parts of the tunnels was therefore subject to uncertainties and it seemed best to limit the number of subcontractors involved in the tunnels. For this reason, many of the subcontracts contained an alternative for the installation stage: either supervision of the installation carried out by a third party or installation by the subcontractors themselves.

Rolling stock policy

13.   The procedure for the rolling stock was slightly different from that used for the fixed equipment. Starting with a definition of performance to be achieved within many tight constraints (space, axle load, environment), the group of contractors, having passed pre-qualification criteria, had to present their technical solutions within the context of a 'request for proposals' (a sort of competition). After examining and adapting these proposals, TML called on the tenderers to supply financial proposals. This procedure led to the selection of the final suppliers.

14.   The method was employed owing to the absence of any comparison for the materials involved.

Fixed equipment

15.   The French Construction Group was responsible for installing the fixed equipment

Fig. 1. Delivery of equipment by the temporary transport system

in the tunnels and at the terminal. This responsibility was conferred to the Tunnel Sub-Project Group for the underground section from the Beussingue Portal, and to the Terminal Sub-Project Group for the above ground section beyond the Portal. Each sub-project group undertook the supervision and the logistics of the installation of the fixed equipment.

Description of equipment

16.   On the surface, as underground, the complex network of fixed equipment injects life into the inert nature of the civil engineering infrastructure (Figs 2 and 3). It consists of the following

(a)   railway tracks on the surface and underground
(b)   electricity supply for trains and auxiliary equipment
(c)   catenaries
(d)   means of control and communication

Fig. 2. Marine service tunnel fixed equipment (typical)

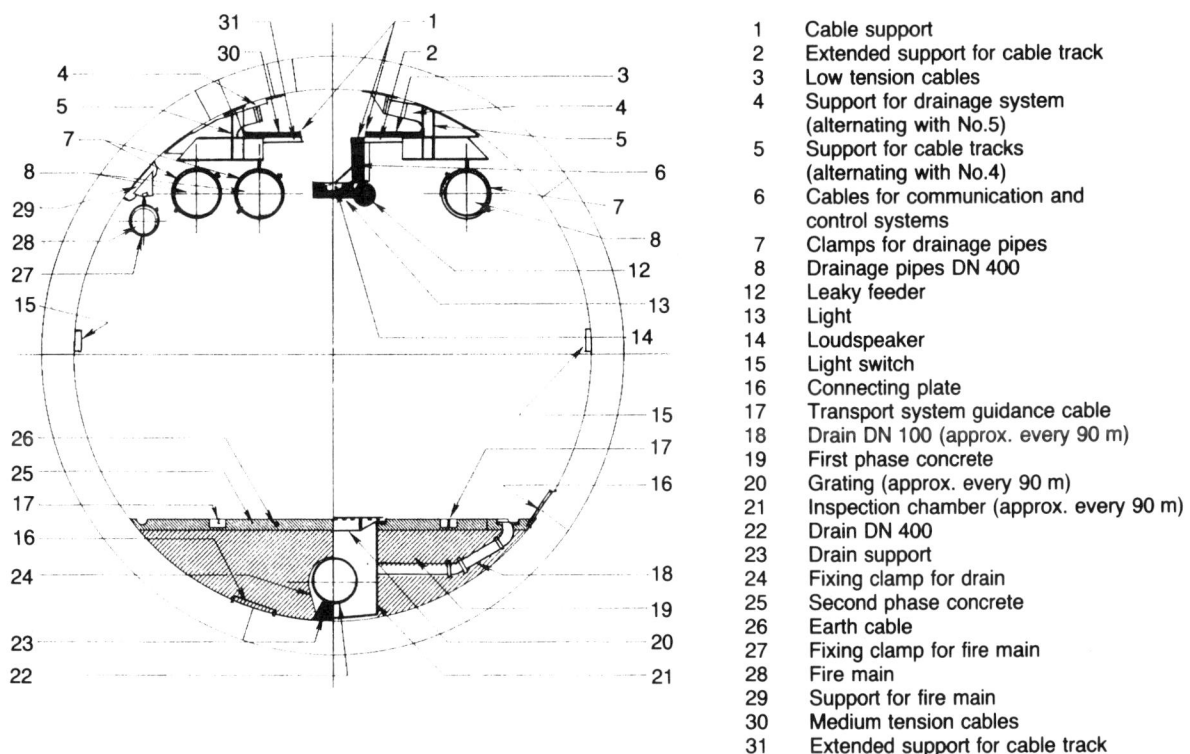

1	Cable support
2	Extended support for cable track
3	Low tension cables
4	Support for drainage system (alternating with No.5)
5	Support for cable tracks (alternating with No.4)
6	Cables for communication and control systems
7	Clamps for drainage pipes
8	Drainage pipes DN 400
12	Leaky feeder
13	Light
14	Loudspeaker
15	Light switch
16	Connecting plate
17	Transport system guidance cable
18	Drain DN 100 (approx. every 90 m)
19	First phase concrete
20	Grating (approx. every 90 m)
21	Inspection chamber (approx. every 90 m)
22	Drain DN 400
23	Drain support
24	Fixing clamp for drain
25	Second phase concrete
26	Earth cable
27	Fixing clamp for fire main
28	Fire main
29	Support for fire main
30	Medium tension cables
31	Extended support for cable track

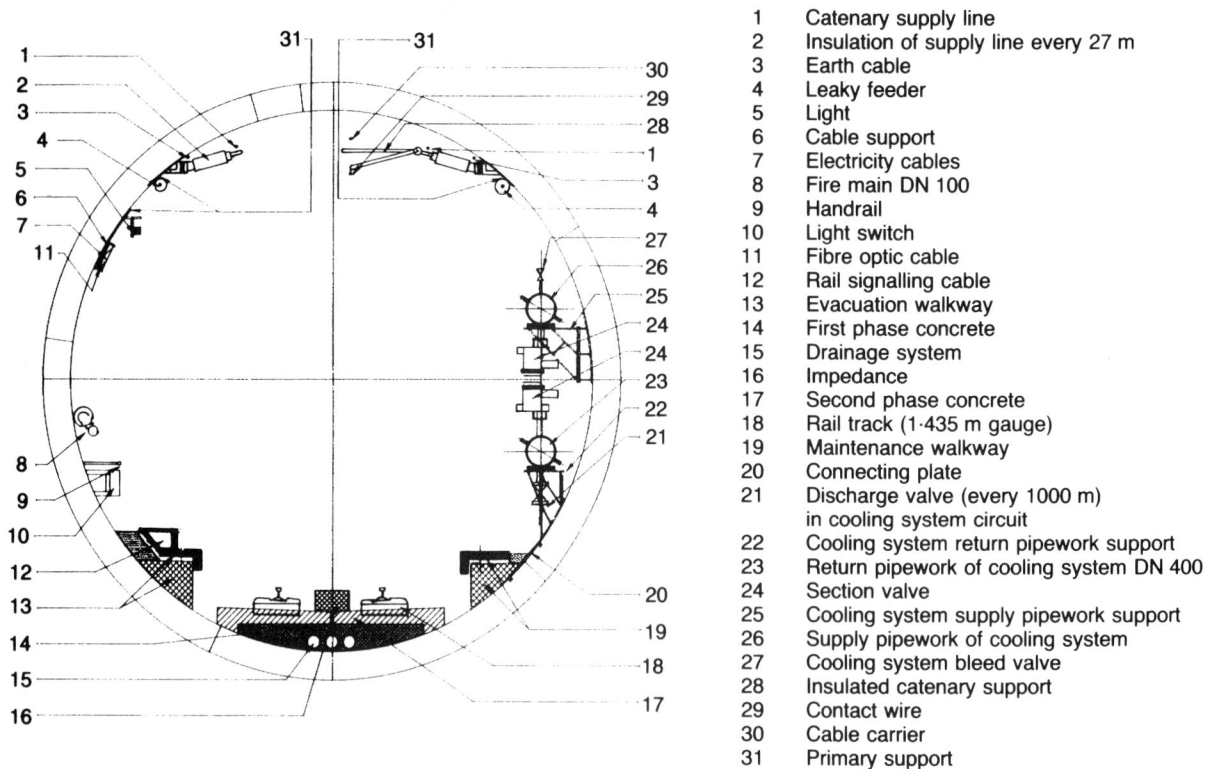

1	Catenary supply line
2	Insulation of supply line every 27 m
3	Earth cable
4	Leaky feeder
5	Light
6	Cable support
7	Electricity cables
8	Fire main DN 100
9	Handrail
10	Light switch
11	Fibre optic cable
12	Rail signalling cable
13	Evacuation walkway
14	First phase concrete
15	Drainage system
16	Impedance
17	Second phase concrete
18	Rail track (1·435 m gauge)
19	Maintenance walkway
20	Connecting plate
21	Discharge valve (every 1000 m) in cooling system circuit
22	Cooling system return pipework support
23	Return pipework of cooling system DN 400
24	Section valve
25	Cooling system supply pipework support
26	Supply pipework of cooling system
27	Cooling system bleed valve
28	Insulated catenary support
29	Contact wire
30	Cable carrier
31	Primary support

Fig. 3. North marine running tunnel fixed equipment

Fig. 4. Rails being placed on track blocks

(e) tunnel ventilation allowing for aerodynamics
(f) drainage of inflows by pumping
(g) means of fire-fighting
(h) tunnel cooling
(i) equipment intended for use in the service tunnel.

The railway tracks

17. The layout and the type of track are designed for shuttle and national train traffic within the limits of the criteria for tonnage, speed and safety defined by contractual requirements (Fig. 4).

Electricity supply

18. The power required for the auxiliary equipment and the trains is supplied by two principal power stations at each terminal and linked to the national 400 000 V networks (Seeboard and Electricité de France) (Fig. 5). Each network supplies half the requirement. In case of failure of either network, the other is capable of supplying all the system, on the condition that they do not have to supply power for non-priority loads and there is a reduction in frequency of shuttles and national trains.

The catenaries

19. Supplied with 25 000 volts, the catenaries are designed to supply the traction power necessary for the shuttles and the national trains in both directions.

20. In the tunnels, within limited space, the overhead power lines consist of a support cable and a contact cable, with longitudinal return conductors fixed to the tunnel walls. The return conductors consist of the rails and two other longitudinal conductors.

21. Sectioning was installed about every 1200 m. Neutral sections are placed along the railway lines in such a way as to avoid contact between two overhead lines whose electrical phasing might be different, and at the mid-point of the tunnels to ensure separation between the British and French networks. Sections allow a faulty section of catenary to be isolated and ensure the resupply of any train after an incident on the catenary.

22. At the terminal, the configuration of the overhead lines is adapted to varying speeds. The type of support is defined by the function of the zones (bridges, platforms, clearance dimensions) (Fig. 6).

Means of control and communication
23. The equipment and the transmission network serves the following functions

(a) railway signalling
(b) management and supervision of the railway system
(c) management and supervision of road traffic
(d) control of auxiliary equipment
(e) telecommunications
(f) data transmission.

All these functions are closely connected and contribute to the running of the whole Transmanche system.

Air supply in the tunnels
24. The ventilation of the tunnels is intended to

(a) renew the air in the tunnels for the safety and comfort of the users; the rate of renewal is based on saturation level (20 000 people), each person needing 26 m³/h of air
(b) ensure the rapid extraction of fumes in the event of a fire in the running tunnels or in the service tunnel
(c) provide a smoke-free refuge for passengers in the service tunnel if the evacuation of a train or shuttle became necessary in the event of a fire
(d) supply fresh air to technical rooms.

Aerodynamics
25. The effect of air piston pressure was one of the matters to be considered when the ventilation system was being designed, in order to provide a satisfactory dynamic stability.
26. The constraints of the varying pressure levels in the running tunnels are attributable to the piston effect of the trains manifested by

(a) resistance to moving locomotives (power requirements)
(b) varying induced pressures which affect the rolling stock, fixed equipment and passengers
(c) loss of energy which reflects on the temperature and humidity in the tunnel
(d) interference with the ventilation system.

27. Piston relief ducts connecting the running tunnels at regular intervals minimize the effects of pressure by reducing the circuit of air between the front and rear of the trains.
28. The spacing (250 m) and the diameter (2 m) of the ducts was designed to reduce the pressure and the speeds of the induced air in

Fig. 5. Principal electricity substation at the Beussingue Portal

Fig. 6. Catenary supports over the French terminal platforms

order to reduce the constraining effects indicated above to satisfactory levels.

Ventilation
29. There are two separate ventilation systems—normal and supplementary—used to control fumes in the running tunnels.

30. *Normal ventilation* The air is supplied from two ventilation plants, one in each shaft at Shakespeare Cliff and Sangatte.
31. The air is directed by way of the shafts into the service tunnel which serves as a duct to distribute the air through cross-passages and grating into the running tunnels. The movement of the air is ensured by the effect of the pressure from the trains. Foul air escapes through the portals.

32. *Supplementary ventilation* Each tunnel has two supplementary ventilation plants, one situated in each shaft at Shakespeare Cliff and Sangatte. The air circulating in the tunnels can be either blown or drawn out from either end, depending on need. Each plant has two reversible ventilation systems (one for each tunnel, each system acting as a back-up for the other).

Tunnel drainage
33. The tunnel drainage system is intended principally to gather and pump out the seepage in the tunnels, but it must also collect other additional water which might come from a rupture in the water supply line, rainwater inflows at the crown of the tunnels, from fire-fighting operations, or from transported goods.

Fig. 7. Tunnel cooling pipes being loaded on to the temporary transport system

34. The water flows under gravity along drainage channels situated in the invert of the running tunnels and in the floor slab of the service tunnel. It is directed into storage tanks (sumps) at the low points of the tunnel, then discharged through three pipelines to the exterior by pumping stations situated close to the sumps.

35. There are five pumping stations in the tunnels: three undersea (one in the French sector and two in the British), and one in each shaft at Shakespeare Cliff and Sangatte.

Fire-fighting measures

36. In addition to the operating policy regulations, the fire prevention requirements are ensured by the use of fire resistant materials and by fire barriers or fire doors which isolate the three tunnels from each other, the technical rooms and the compartments.

37. Smoke detectors are installed in all the technical rooms, with automatic extinguishing devices (by halon gas) and remote control systems which enable the ventilation to be cut off, damper vents to be closed and service personnel to be alerted.

38. A dedicated water supply line in the service tunnel, divided into sections (land and marine on each side), feeds the fire hydrants in each cross-passage and in the running tunnels.

39. Each section is fed by an 800 m³ storage tank and a pumping station at the two portals and in each shaft. The supplies are interconnected to cater for failure in the system in a section.

Tunnel cooling

40. In the running tunnels, the air friction induced by the trains raises the temperature considerably. Calculations demonstrate that neither the ventilation nor the natural dissipation (ground, seepage water) are sufficient to maintain the tunnels at an acceptable temperature of about 25°C.

41. Each tunnel has been divided into four sectors, namely land and marine, UK and French. These are cooled separately by the circulation of refrigerated water in each section through a discharge pipe and a return pipe (Fig. 7).

42. Chilling and circulation of refrigerated water is provided by two plants near each shaft. Each plant covers the needs of cooling two sections under sea and under land in each tunnel. The water in the pipes may be just at a low temperature or it may be a mixture of ice and water.

Technical rooms

43. The technical rooms contain electrical and mechanical equipment required by the auxiliary services in the tunnels. They are spaced at regular intervals on each side of the service tunnel and are similar to cross-passages except that the running tunnel end is permanently closed off.

44. Electricity substations fed with high voltage provide medium and low voltage supplies and are situated 3–5 km apart. Electrical equipment rooms supply the different voltages needed for the operation and control of the auxiliary equipment and are placed at 750 m intervals.

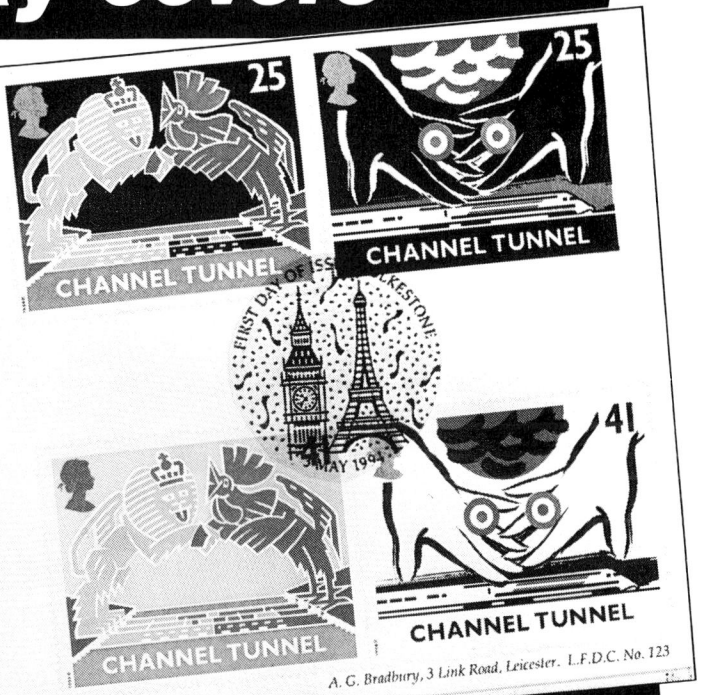